ENVIRONMENTAL PHYSIOLOGY
OF DESERT ORGANISMS

ENVIRONMENTAL PHYSIOLOGY OF DESERT ORGANISMS

Edited by

NEIL F. HADLEY
Arizona State University

 Dowden, Hutchinson & Ross, Inc.

STROUDSBURG, PENNSYLVANIA

Distributed by
HALSTED   A divison of
PRESS   John Wiley & Sons, Inc.

LIBRARY OF CONGRESS CATALOGING IN PUBLICATION DATA
Main entry under title:

Environmental physiology of desert organisms.

   Proceedings of a symposium held at Arizona State University, Tempe, 1974, sponsored by the American Institute of Biological Sciences and others.
   1. Adaptation (Physiology)--Congresses. 2. Desert biology--Congresses. I. Hadley, Neil F. II. American Institute of Biological Sciences. [DNLM: 1. Ecology--Congresses. 2. Environment--Congresses.
3. Desert climate--Congresses. QH541.5.D4 E61 1974]
QP82.E6   574.5'265   75-14408
ISBN 0-470-33851-2

Exclusive distributor: **Halsted Press**
A Division of John Wiley & Sons, Inc.

PREFACE

The invitation to organize a symposium on "Environmental Physiology of Desert Organisms" to be held in conjunction with the 1974 American Institute of Biological Sciences meetings was greeted with both enthusiasm and reservation. I agreed that Arizona State University, Tempe, Arizona, would be an appropriate setting for such a symposium, for the Sonoran Desert which surrounds the campus had been the scene for many classic studies in desert ecophysiology. More important, it would enable those in attendance to experience first-hand the nature of the desert environment and to fully appreciate adaptations possessed by its inhabitants. I was hesitant, however, in that many excellent reviews and symposia, some quite recent, already existed on this general topic. The time limitations imposed on the proposed symposium would make it impossible to cover all subjects warranting attention under such an ambitious title, nor would it be possible to include in the program many individuals who had contributed significantly to this research area.

Plans for the Tempe symposium were initiated with hopes of at least partially overcoming these problems. Participants and topics were selected to provide an interdisciplinary coverage of principles and concepts underlying the adaptational biology of desert organisms, with emphasis placed on groups or topics not previously subjected to review. Furthermore, emphasis was to be on studies that reflected application of innovative techniques in their research design. Although general topics were more-or-less dictated to symposium participants, the latter were given total flexibility in their approach and coverage.

The symposium was divided into three sections: desert species and dry heat; adaptations at the cellular and molecular levels; and desert resources and species requirements. General topics included in the first section were the energy flux and thermal regime of desert organisms, their morphological and physiological responses to extreme temperature and drought, and water turnover and seasonal changes in tissue fluids. The second section examined enzymatic compensation and photosynthetic adaptations to thermal stress, cycling of carbon dioxide and acclimatization in gas exchange by desert plants, and roles of the endocrine and cardiovascular systems in animals adapted to arid environments. The final section included adaptive mechanisms exhibited by temporary pond inhabitants, recent developments in techniques used to monitor physiological responses in free-ranging animals, and the future of desert ecosystems in light of man's increasing activities in arid regions. These sections and the sequence in which papers were presented during the symposium have been retained in this publication.

The symposium was initially organized by AIBS and the Physiological-Ecology Section of the Ecological Society of America, and was subsequently cosponsored by the American Society of Zoologists and the US/IBP Desert Biome. Their generous financial support for costs associated with the symposium's presentation and the publication of the proceedings is gratefully acknowledged. I thank our publishers, Dowden, Hutchinson & Ross, Inc., in particular James B. Ross, for their continued interest and cooperation in all aspects of the symposium.

The following individuals helped ensure quality and accuracy of contributed papers through their critical review of the manuscripts: Jerome Aronson, Joseph Bagnara, Michael Barbour, Gordon Bender, Robert Chew, Eric Edney, Kenneth Fry, Francis Horne, Sherwood Idzo, Bruce Kimball, John Minnich, Robert Mullen, David Rasmussen, William Stone, Boyd Strain, and Walter Whitford.

A special thanks is extended to the late Director of AIBS, John R. Olive. His enthusiastic support during the initial planning stages and fruitful efforts to obtain travel funds for foreign participants were instrumental in guaranteeing the success of the symposium. Dedication of this publication in his memory cannot adequately express my sincere appreciation for his efforts on my behalf.

NEIL F. HADLEY

CONTENTS

ENVIRONMENTAL PHYSIOLOGY
OF DESERT ORGANISMS

HEAT FLUX AND THE THERMAL REGIME OF DESERT PLANTS

Duncan T. Patten and Edward M. Smith

*Department of Botany and Microbiology, Arizona
State University, Tempe, Arizona*

Abstract

The thermal regime of a plant is a result of both biotic and abiotic factors. The basic concepts of energy budgets for organisms have been discussed by many researchers. In addition to plant physiology and morphology, they all emphasize the factors of solar radiation, environmental temperatures, wind, and available moisture for transpiration as controlling factors of the plant energy balance.

In this paper we compare the abiotic factors of air and soil temperature, and solar radiation and wind between the cold and hot North American deserts, and examine their influence on various thermal macroenvironments of each desert.

The Sonoran Desert is selected to show variations in some abiotic factors at specific sites and microsites. Solar radiation input is related to cloudy or potentially rainy periods, showing greater fluctuations during these periods than during dry seasons.

The influence of plants on the thermal microenvironment of other plants is shown. Small trees, such as *Cercidium microphyllum,* create moderated environments by influencing radiation input. Smaller shrubs do not have as much influence.

Comparisons are made between leaf and stem temperatures of various desert shrubs and succulents. Temperatures of shrubs are only a few degrees different from ambient, with larger, more mesic shrubs such as *Ambrosia* being warmer and small-leaved shrubs such as *Larrea* being cooler than ambient under warm conditions. Moisture stress differentials between these species is suggested as a possible explanation, along with leaf size and orientation.

Cacti experience temperature buildups as much as 15°C above ambient even under warm conditions, depending on morphology and orientation. They rapidly cool when the sun goes down, suggesting the importance of nighttime transpiration as a cooling mechanism.

The variation in desert plants, including shrubs, succulents, and mesophytic annuals, parallels the variation in thermal microsites found in the desert.

INTRODUCTION

The desert environment is often considered to be the harshest of any biome. Biological functions in general are dependent on moisture and tolerate a relatively narrow temperature range. Long exposures below freezing or above 45 to 50°C may be lethal to many organisms. Desert organisms are afforded two possibilities for survival. They either can live with the environment on the surface, being morphologically adapted to take advantage of the more moderate environments the desert may offer during certain seasons and to tolerate extremes at other times, or they can avoid extremes by moving to more moderate microenvironments.

Although this paper is primarily concerned with the environment of plants, this environment is similar to that experienced by many desert animals. The total ecological community is a function of and influences its own environment; this paper presents a broad picture of the variations in desert thermal environments and will provide a common background on thermal regimes of desert organisms for later physiological discussions.

ENERGY-EXCHANGE CONCEPTS

The primary source of energy in a biological system is solar radiation. Upon entering the ecosystem, it is reflected, absorbed, and dissipated, creating a balance between energy input and outflow. For any surface, whether it be plant leaf, rock, or animal, there is a basic energy balance represented by the following equation (Gates, 1962):

$$S \pm R \pm LE \pm C \pm G \pm s = 0 \qquad (1)$$

where S = shortwave radiation (solar and scattered sky), R = thermal or longwave radiation (terrestrial and atmospheric), LE = energy flux due to latent heat of evapotranspiration or condensation, C = convection, G = conduction, and s = short-term energy storage or release. Energy flux due to metabolism is small and is omitted from the equation.

Important factors in this equation are the solar energy input (S), the thermal regime (R), and factors related to wind movement, that is, LE

and *C*. All these factors are readily measured directly or indirectly and indicate the potential extremes of the thermal regime of an organism.

Longwave radiation (*R*) in the environment is a result of the temperature of the atmosphere and all objects in the system. These objects radiate longwave radiation based on the following equation:

$$R = \varepsilon\sigma T^4 \tag{2}$$

where *T* = absolute temperature of the radiating surface, σ = Stefan–Boltzmann constant, and ε = emissivity of the object (1.00 equals a perfect radiator). The cloudless atmosphere radiates according to the equation

$$R_A = \sigma T^4 \{1 - c \exp[-d(273 - T)^2]\} \tag{3}$$

where *c* = 0.261 and *d* = 7.77 $\times$ 10^{-4} (Idso and Jackson, 1969). From these equations one can see how important temperatures of air, soil, and other inanimate objects are to the thermal environment of a plant.

The temperature of a plant leaf or other plant part is thus dependent, in part, on its heat load. This is basically the sum of shortwave and longwave radiation in the environment expressed as (Gates, 1963)

$$H = (1 + r)(S + s) + R_G + R_A \tag{4}$$

where *H* = total radiant energy load on a horizontal surface, *S* + *s* = direct and scattered solar radiation, *r*(*S* + *s*) = reflected solar radiation, R_G = thermal radiation from the ground [see equation (2)], and R_A = thermal radiation from the air [see equation (3)].

To this point, the thermal regime of a plant and potential plant temperatures have been related only to abiotic factors. Obviously, the morphology of the organism also plays a role in energy absorption and reradiation and thus organism temperature.

Desert organisms have adapted in many ways to the great variance in temperature and moisture found in the desert (Hadley, 1972). Some plant species evolved to meet desert conditions head on. Cacti are good examples in that they have over time lost the primary water loser, the leaves, and evolved a sink for moisture, the stem. Cacti and other succulents are susceptible to temperature buildup. As a result, their metabolic functions appear to be heat tolerant. Spines of many cacti also reflect light and, hence, delay or reduce diurnal temperature buildup (Gibbs and Patten, 1970).

Other desert plants avoid extreme temperatures by occurring only under moderate temperature conditions and optimum moisture conditions. Winter desert annuals best exemplify this type of plant. They grow during milder climatic times and often do best in temperature and moisture environments created by shrubs and small trees (Patten and

Smith, 1973). These microsites under shrubs are, in themselves, interesting locations for desert thermal flux studies.

The majority of desert plants are shrubs or small trees. These plants have leaves of various size and longevity, deciduousness being another mechanism used by desert plants to avoid drought conditions. Plant temperatures in the desert and elsewhere are closely related to the plant's ability to carry on transpiration (Gates, 1964; Lange 1965; Idso and Baker, 1967). Desert shrubs that are deciduous only under extreme drought conditions, such as *Ambrosia deltoidea,* or appear to be evergreen, such as *Larrea divaricata,* respond differently to the desert thermal regime than do plants that have leaves only when water for transpiration is readily available.

Leaf temperatures are closely related to the size and orientation of the leaf in addition to the transpiration rates (Gates et al., 1968). The equation

$$Q_{abs} = \varepsilon\sigma T_\ell^4 \pm k_1\left(\frac{V}{D}\right)^{\frac{1}{2}}(T_\ell - T_a)$$
$$+ L\frac{s\rho\ell(T_\ell) - \text{r.h.}\cdot s\rho a(T_a)}{r_\ell + k_2(W^{0.20}D^{0.35}/V^{0.55})} \tag{5}$$

as described in Gates et al. (1968) emphasizes radiation input [light and thermal (Q_{abs})], air temperature (T_a), thermal radiation from the leaf $(\varepsilon\sigma T_\ell^4)$, and wind (V), and also recognizes factors that influence plant water content and play a role in the energy budget of the leaf. These include the moisture factors relative humidity (r.h.), the density of water vapor at saturation within the mesophyll at leaf temperature $[s\rho a(T_\ell)]$, and the internal diffusion resistance to water vapor of the leaf (r_ℓ). The importance of leaf size and orientation is expressed in the terms W, the leaf width, and D, the dimension of the leaf in the direction of the wind. From this equation it is clear that a leaf without much width, but with maximum orientation to the wind, might better dissipate absorbed energy than a thick leaf or plant part that is not oriented for maximum exposure to the wind.

To give examples of the relative significance of leaf size and internal diffusion resistance (r_ℓ) differences, and wind velocity, air temperature, and solar radiation changes on differences between air and leaf temperature, comparisons have been made from equations and figures in Gates (1965) and Gates et al. (1968).

1. At an air temperature (T_a) of 30°C, radiation of 1.2 cal cm^{-2} min^{-1}, r_ℓ = 50 s cm^{-1}, and wind velocity equaling 100 cm^{-1}, leaf temperature minus air temperature (ΔT) equals 3°C for a 1 × 1 cm leaf and 13°C for a 10 × 10 cm leaf.

2. For a small leaf (1 × 1 cm), typical of many desert plants, with

radiation, leaf resistance, and air temperature the same as in example 1, ΔT is 6°C for a wind velocity of 223 cm s^{-1} (5 mph), 10°C for a wind velocity of 45 cm^{-1} (1 mph), and 25°C for no air movement and only free convection.

3. If conditions were the same as in example 2 and air temperature raised to 40°C, ΔT for a 1 × 1 cm leaf would be 5, 8.5, and 22°C at wind velocities of 223, 45, and 0 cm s^{-1}, respectively.

4. For a 1 × 1 cm leaf with low leaf resistance, a radiation input of 1.3 cal cm^{-2} min $^{-1}$, and a wind velocity of 45 cm^{-1}, ΔT changes from 10 to 16°C when T_a changes from 40 to 10°C; when wind velocity equals 223 cm^{-1}, ΔT changes from 5.8 to 8.2°C when T_a changes from 40 to 10°C.

5. Under abiotic conditions of air temperature at 30°C, radiation equaling 1.2 cal cm^{-2} min^{-1}, wind velocity at 100 cm s^{-1}, and relative humidity at 20 percent, ΔT changes from 2.5 to 3°C for a 1 × 1 cm leaf when r_ℓ changes from 10 to 50 s cm^{-1}; for a larger leaf (10 × 10 cm), ΔT changes from 10 to 12.5°C for an r_ℓ change from 10 to 50 s cm^{-1}.

6. For a 1 × 1 cm leaf in a wind velocity of 223 cm s^{-1}, where $T_a = $ 10°C, ΔT is 3°C for a radiation input of 0.8 cal cm^{-2} min^{-1} and 8.2°C for radiation at 1.3 cal cm^{-2} min^{-1}. Where $T_a = $ 40°C, ΔT is 0.5 and 5°C at radiation inputs of 0.8 and 1.3 cal cm^{-2} min^{-1}, respectively.

These examples illustrate that (1) changes in wind velocity influence leaf temperatures more than air temperature changes, (2) leaf size is closely related to differences in leaf temperatures and is definitely more important than internal leaf diffusion resistance, and (3) changes in both air temperature and radiation input have a moderate influence on differences between leaf temperature and air temperature, with normal decreases in radiation probably being more significant than decreases in air temperature.

ABIOTIC FACTORS IN NORTH AMERICAN DESERTS

The abiotic data presented are from U.S. International Biological Program desert biome validation or process studies, which were selected because they represent data from "typical" desert communities in North America. These include the Great Basin Desert (or cold desert), and the Mohave, Sonoran, and Chihuahuan deserts (or hot deserts) (Shreve, 1942). All locations from which data were obtained are not centrally located within each major desert, but they typify the desert in the general region of the sample site.

Figure 1 represents monthly average maximum and minimum air temperatures in the four North American deserts. The Sonoran Desert is generally the hottest with summer average maximum temperatures exceeding 40°C. The Mohave Desert temperatures are not as warm, quite

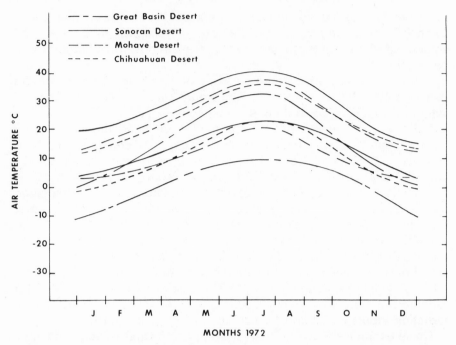

FIGURE 1 Average monthly maximum and minimum air temperatures for 1972 at the four US/IBP desert biome validation sites. (Based on data from Balph, 1973; Thames, 1973; Turner, 1973; Whitford, 1973.)

possibly because the Mohave site is on the northern edge of this desert. The Great Basin Desert, although considerably colder than the other deserts in winter, approaches the maximum temperatures of the hot deserts in summer. The Great Basin also has the widest range between summer average maximum and minimum air temperatures and summer and winter maximum temperatures of all the deserts. This greater temperature range in the Great Basin indicates a potentially harsher environment in terms of heat flux and plant adaptation to temperature changes.

The major temperature fluctuations in a system are a response to fluctuations in solar radiation input. Figure 2 shows curves drawn generally through the maximum solar input points for three of the deserts. Solar energy maxima, which correspond to air temperature data, are highest for the Sonoran Desert. The Great Basin Desert is lowest throughout the year. Although summer maxima in the Great Basin do not equal those of the hot deserts, the more northern latitude of the Great Basin site causes longer periods of solar energy input; thus, summer daily totals may well exceed those of the hot deserts. Winter solar input in the Great Basin is less than half that in hot deserts, which

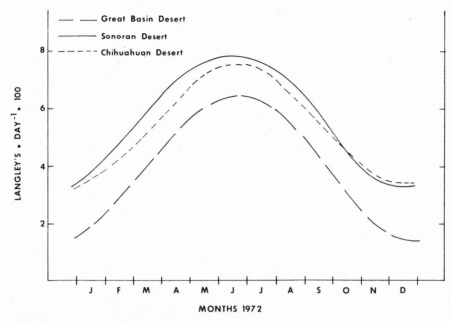

FIGURE 2 Curves drawn through the majority of maximum daily total Langley points for 1972 comparing three North American deserts. (Based on data from Balph, 1973; Thames, 1973; Whitford, 1973.)

explains why hot desert minimum air temperatures generally stay above freezing (Fig. 1).

Soil temperatures can be used as an indication of total energy input because soil acts as a heat sink in summer. Although data are not available for the same soil depths from all US/IBP desert biome sites, average monthly maximum and minimum soil temperatures for near-surface levels do show some trends (Fig. 3). As expected from the solar radiation input (Fig. 2), the Sonoran Desert has the hottest near-surface soils of the three deserts presented. Soil from the Mohave Desert may be cooler because measurements were taken 1.5 cm deeper than at the Sonoran site. As with air temperatures (Fig. 1), the Great Basin Desert shows the greatest seasonal soil temperature fluctuations, because the more northern latitude of the site, as compared to the hot desert sites, causes longer total daily radiation input and because measurements were taken at or near the soil surface. Again, soil temperature fluctuations of the Great Basin demonstrate a potentially harsher annual environment than the hot deserts, although average maximum soil temperatures over 60°C as measured in the Sonoran Desert also must be considered extreme.

To some extent, wind influences leaf temperatures of plants. All deserts are quite breezy (Fig. 4). No desert is much windier than the others,

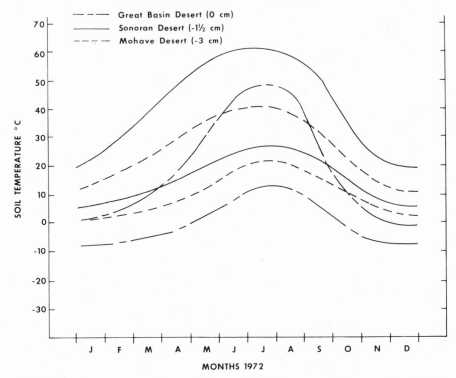

FIGURE 3 Average monthly maximum and minimum soil (near-surface) temperatures for 1972 at three US/IBP desert biome validation sites. (Based on data from Balph, 1973; Thames, 1973; Turner, 1973.)

although the Sonoran Desert had (1972) one unusually windy month. Average wind velocities of between 6 and 9 km·h^{-1}, which occur primarily during the day, aid in convective loss of heat from plants and maintain high transpiration rates when water is plentiful. Wind movement in the desert, however, often occurs in gusts with lulls in between. Thus, total wind averages only give an indication of the potential convective cooling within the various desert ecosystems.

ABIOTIC FACTORS IN THE SONORAN DESERT

The US/IBP Sonoran Desert location was shown in earlier figures to be the hottest of the North American desert sites. It also apparently has greater radiation peaks than the other deserts (Fig. 2). Therefore, we selected the Sonoran Desert for more detailed examination of radiation and temperature variations that must be endured by plants.

Figure 5 shows the range in maximum peak radiation input experienced at one Sonoran Desert site. The greatest fluctuations occur during rainy or potential rainy seasons. Normally, July through August and

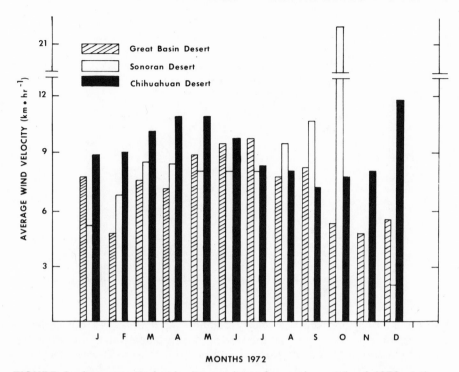

FIGURE 4 Average wind velocity per hour for each month of 1972 at three US/IBP desert biome validation sites. (Based on data from Balph, 1973; Thames, 1973; Whitford, 1973.)

December through March are the "typical" rainy seasons. These are, at least, the times of year when there is most cloud cover. Maximum daily radiation during these cloudy periods may drop as low as 0.25 cal cm^{-2} min^{-1}; however, when cloud cover is scattered, the radiation peak from direct solar and scattered radiation may approach 2.00 cal cm^{-2} min^{-1}. During late spring when skies are usually clear, the variation in maximum radiation input may range only from about 1.30 to 1.60 cal cm^{-2} min^{-1}.

Influence of Plants on the Thermal Environment

Plants in the Sonoran Desert may appear to be evenly spaced with little influence on each other; however, this is generally not the case. Small shrubs may grow under larger shrubs or small trees such as palo verde (*Cercidium* spp.). In addition, during rainy seasons annual plant growth is much greater under the canopy of the shrubs and trees (Smith and Patten, 1974). For this reason a discussion of the thermal regime of plants would not be complete unless comparisons were made of the various microsites in which plants control, in part, the environment.

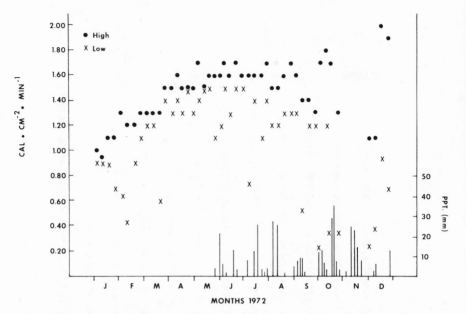

FIGURE 5 High and low maximum radiation input per week compared to precipitation periods for 1972 at the US/IBP desert biome validation site in the Sonoran Desert near Tucson, Arizona. (Based on data from Thames, 1973.)

Two common shrubs or small trees in the Sonoran Desert are *Cercidium microphyllum* (palo verde) and *Larrea divaricata* (creosote bush). The canopies of these shrubs shade the ground and create moderated environments. Measurements of solar radiation taken under their canopies with a Belfort pyreheliograph show that *Cercidium* reduces incoming radiation considerably more than does *Larrea* (Fig. 6). Solar radiation measured under the north aspect canopy of *Larrea* averaged 81 percent of that measured in the open; averages were 43 and 38 percent of the radiation in the open under the south and north aspect canopies of *Cercidium,* respectively. The potential heat load in all three microsites would be reduced for any other plants growing under canopies of the shrubs.

The heat load would not only be reduced directly for plants under shrub canopies but would be reduced indirectly through reduction of air and soil temperatures in this microhabitat. On a typical day in March in the Sonoran Desert when air temperatures in the open approach 30°C, soil and air temperatures near the soil surface in the open and under *Cercidium* are quite different (Fig. 7). The soil temperature at -1.5 cm in the open may be 25°C above air temperature (15 cm), while soil temperatures at the same depth under *Cercidium* do not exceed air temperature. Air temperatures in the open at 1.5 cm may be 20°C higher than ambient temperatures at 15 cm, while temperatures under the

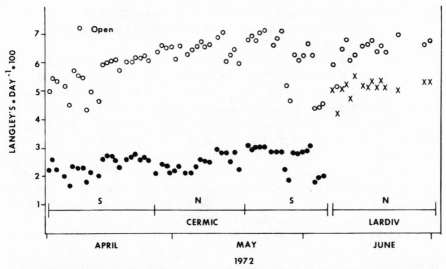

FIGURE 6 Total solar radiation input per day in the open and under the north and south aspect canopies of *Cercidium microphyllum* (CERMIC) (dots) and the north aspect canopy of *Larrea divaricata* (LARDIV) (×'s) at a US/IBP process study site in the Sonoran Desert near Cave Creek, Arizona.

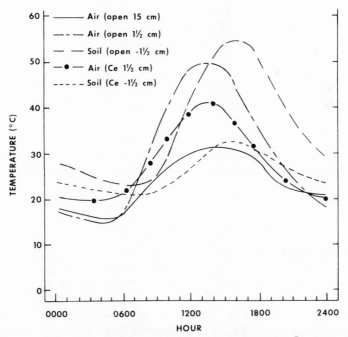

FIGURE 7 Diurnal air and soil temperature fluctuations in the open and under *Cercidium microphyllum* (Ce) for a day in March 1973 at a US/IBP process study site in the Sonoran Desert near Cave Creek, Arizona.

canopy are only 10°C higher. These moderations in air and soil temperatures under the shrub canopy are obviously a result of reduction in total radiation input (Fig. 6).

Seasonal differences in air and soil temperatures in the open compared to under the canopy of *Cercidium* show a continual moderation of the latter microenvironment (Figs. 8 and 9). During the winter growing season (December to March), maximum air temperatures (1.5 cm) remain about 10°C lower under the canopy than in the open (Fig. 8). Minimum air temperatures are not as well moderated by the canopy. One week of air-temperature measurements under the south aspect canopy of *Ambrosia deltoidea* showed little difference from air temperatures in the open. Thus, canopies on exposed sides of small shrubs have little influence on microsite air temperature but do influence subsurface soil temperature (see *Ambrosia,* in Fig. 9).

Subsurface soil temperatures (−1.5 cm) under *Cercidium* are moderated more than air temperatures (Fig. 9). When maximum air temperatures (1.5 cm) were reduced by 10°C under the canopy, subsurface soil temperatures were 15 to 20°C lower under *Cercidium* than in the open. These reduced temperatures obviously lower the heat load on

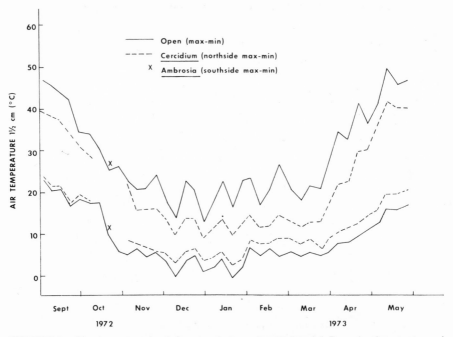

FIGURE 8   Maximum and minimum air temperatures at 1.5 cm in the open and under shrubs during the period encompassing the 1972–1973 winter growing season at a US/IBP desert biome process study site in the Sonoran Desert near Cave Creek, Arizona.

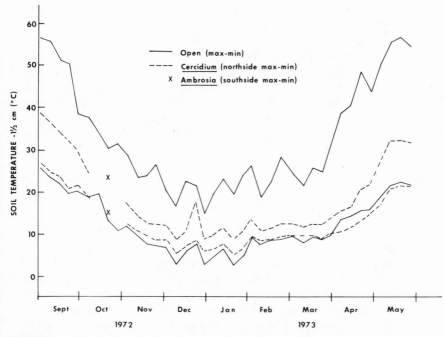

FIGURE 9   Maximum and minimum soil temperature at −1.5 cm in the open and under shrubs during the period encompassing the 1972–1973 winter growing season at at US/IBP desert biome process study site in the Sonoran Desert near Cave Creek, Arizona.

plants under the canopy and reduce the evaporative demands; thus, soil moisture is maintained, which permits uninterrupted transpiration for greater cooling of the leaves of annual plants that commonly thrive in these microenvironments.

PLANT TEMPERATURES IN THE DESERT

Lange (1965) discusses the factors considered by Huber (1956) and Raschke (1956) to be most important in determining leaf temperature. These are radiation balance (i.e., the difference between adsorbed radiation and longwave reradiation), transpiration rate and the latent heat of vaporization at a particular leaf temperature, convective heat exchange with the environment, chemical energy exchange (relatively insignificant), and heat exchange due to water transport. These are basically the same factors presented by Gates et al. (1968), although they also emphasize leaf width and leaf orientation.

Both Lange (1965) and Gates et al. (1968) present leaf temperature data, but Lange's were taken with probes (thermister for fleshy leaves and thermocouples for thin leaves), whereas Gates took temperatures with a pistol-grip infrared radiometer. Gibbs and Patten (1970) used thermocouples for measurements in both thin leaves and succulents. All three papers include examples of desert plants that have leaf temperatures above the surrounding air temperature as well as plants that are capable of maintaining leaf temperatures below ambient. Lange (1965) showed that a species of *Zygophyllum* had leaf temperatures 10°C above ambient (30°C); another arid land species, *Citrullus cologynthis,* maintained leaf temperatures 10°C below an ambient temperature of 50°C. When a *Citrullus* leaf was excised and transpiration stopped, the leaf temperature increased to 60°C.

Gibbs and Patten (1970) reported that Sonoran Desert shrub leaves generally had temperatures above ambient when ambient was cool (ca. 20°C). Leaf temperatures varied with species at warmer ambient temperatures, some species being capable of maintaining leaf temperature below high (above 30°C) ambient temperatures.

Leaf temperatures of all species measured by Gates et al. (1968) in the southern part of the Great Basin Desert were a few degrees Celsius above ambient some time during the day. These measurements were made in August when maximum air temperatures were between 30 and 33°C. The one exception was an *Opuntia* species that had stem (not leaf) temperatures 10 to 16°C above ambient. Gibbs and Patten (1970) also showed much higher stem succulent temperatures in comparison to normal leaf temperatures. They also demonstrated the influence of exposure and shape of stem succulents on diurnal stem temperature patterns.

Figure 10 shows how two different Sonoran Desert shrubs respond to maximum ambient temperatures above 30°C in June. *Larrea* has small leaves, often with limited exposure to the sun, whereas *Ambrosia* has larger leaves with surfaces exposed to the sun.

The fact that *Larrea* can maintain leaf temperatures below ambient when *Ambrosia* cannot may be closely correlated to the diurnal response of the two species to moisture stress (Halvorson and Patten, 1974). *Ambrosia* develops a much lower stem water potential (measured by a pressure bomb) during the middle of the day than does *Larrea* under similar conditions. This low water potential, at times below −80 bars, may reduce transpiration enough that when the leaves are fully exposed to solar radiation they cannot cool below ambient. The late afternoon rise of *Larrea* leaf temperature above ambient may also be related to internal water stress, low transpiration, and greater leaf exposure. It is also possible that the higher leaf temperatures caused higher rates of plant water loss and, thus, decreases in stem water potential.

In an attempt to duplicate some of the diurnal temperature patterns found in *Opuntia englemannii* (now *O. phaeacantha*) by Gibbs and

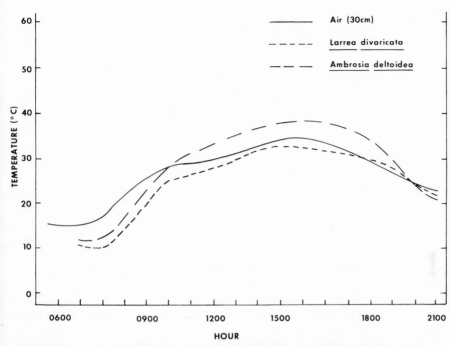

FIGURE 10   Average temperatures during three sunny days with slight breezes in June of the air at 30 cm and leaves of *Larrea divaricata* and *Ambrosia deltoidea*.

Patten (1970), fine wire thermocouples were placed near the center of pads of *Opuntia basilaris,* another species of prickly pear. Pads were selected for exposures with sides facing N–S, E–W, SE–NW, and SW–NE. All pads heated above ambient air temperature (Fig. 11); however, the E–W facing pad cooled down almost to air temperature at midday when only the edge of the pad was exposed to the sun. In other pads, temperatures rose more than 15°C above ambient. If the succulent pads were not so massive, they probably would not heat more than a few degrees above ambient under optimal moisture conditions, because, when moisture is not limiting, there may be midday stomatal opening and transpiration.

The stem pads of prickly pear cactus species respond to the thermal environment much like extra thick leaves, holding heat and readily dissipating it through reradiation and transpiration at night, thus cooling well below ambient air temperatures. However, stem succulents that are more cylindrical in shape have a greater potential heat storage than do pad-shaped stems.

Gibbs and Patten (1970) showed some of the temperature differences in the center of *Echinocereus engelmannii* (hedgehog cactus) from the

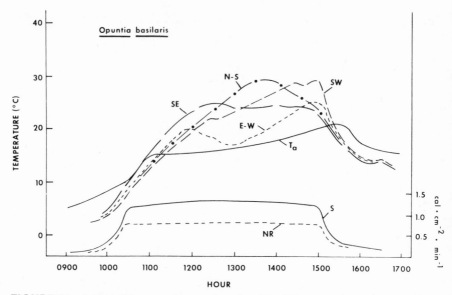

FIGURE 11 Internal temperatures of pads with different orientation on a beaver tail cactus (*Opuntia basilaris*) in January in relation to the surrounding air temperature (T_a), solar radiation (*S*), and net radiation (*NR*). The study site was located in such a way that early morning and late afternoon direct solar radiation were blocked and the wind velocity averaged about 415 cm·s^{-1}.

top down, and how solar radiation and soil temperatures caused differential heating on both ends of the stem. Temperatures were taken with thermocouple probes placed 2 mm under the cuticle on the exposed side and shaded side of stems of different species of *Echinocereus* in the Pinaleño Mountains in eastern Arizona (Fig. 12). Although these species are slightly different in morphology and grow at quite different elevations, ranging from 1260 to 2500 m, they absorb solar radiation and respond to ambient air temperature in much the same manner. The shaded sides never rose more than a few degrees above ambient, while the exposed sides were 10 to 15°C above ambient. Like the prickly pear cactus, when the exposed side no longer was radiated by the sun, the tissue temperature at the 2-mm depth rapidly dropped to near ambient and probably dropped below, but later measurements were not taken.

DISCUSSION

The thermal environment in deserts is determined by season, abiotic factors such as temperature, radiation, wind, and moisture, and biotically controlled factors such as shrub canopy growth. Microenvironments for plants may have temperatures exceeding 60°C with high solar

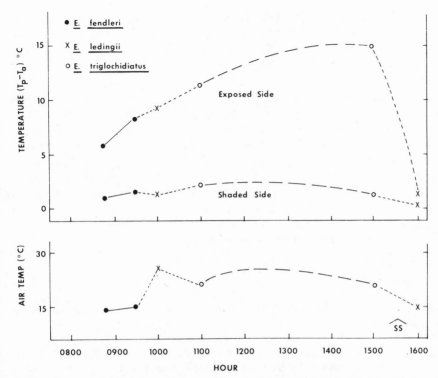

FIGURE 12 Temperature differences between the stem temperature (exposed and shaded sides) (T_p) and air temperature (T_a) of three species of *Echinocereus* (hedgehog cactus) in the Pinaleño Mountains, Arizona, in September. *Echinocereus fendleri* occurs at 1260 m, *E. ledingii* at 1670 m, and *E. triglochidiatus* at 2500 m. Sunset (SS) occurred in the *E. ledingii* area on the east side of the mountain at about 1530 hours. (Based on data from Dinger, 1971.)

radiation input or temperatures not much above 20°C with low solar radiation. Each site demands a different physiological response of the plants that occupy it. Plant temperatures, which do not always follow air temperatures, are dependent on the plant's morphology, exposure orientation, and capabilities to prevent heat absorption through reflection and to dissipate heat through transpiration and convection.

In the most exposed, harsh sites, such as in the open on a south-facing slope in the Sonoran Desert, plants such as *Opuntia bigelovii* (teddy bear cholla cactus) dominate. They can withstand high temperatures, and maintain a moisture balance due, in part, to high reflectivity (Gibbs and Patten, 1970). The exposure on north-facing slopes is not as great, and soil and near-ground temperatures are moderate through most of the winter growing season (Patten, 1973). These slopes are dominated by *Ambrosia deltoidea*. *Larrea* is common on either slope.

Those desert plants that function more like mesophytes than zerophytes (i.e., annuals) grow best in a moderated microenvironment (Smith and Patten, 1974). The shade formed by palo verde trees (*Cercidium*) is an example of a biotically controlled desert thermal environment. In this environment, the large-leaved annuals have low moisture stress, high transpiration, and low heat load, all factors that keep leaf temperatures well below their lethal point and near air temperature.

Except for plants like the annuals, most desert plants survive under the thermal regime of the desert by maintaining an energy balance through nighttime cooling and by being physiologically adjusted to function at high temperatures. However, in the desert when temperatures become too high for survival because moisture is not available for transpirational cooling, the leaves and small branches die back, enabling the plant to conserve moisture and stored energy. Conditions in the desert will always improve another day.

LITERATURE CITED

Balph, D. F. (coordinator). 1973. Curlew Valley validation site report. US/IBP Desert Biome Research Memorandum 73-1. 333p.

Dinger, B. E. 1971. Comparative physiological ecology of the genus *Echinocereus* (Cactaceae) in the Pinaleño Mountains, Arizona. Ph.D. Dissertation, Arizona State University, Tempe, Ariz. 111 p.

Gates, D. M. 1962. Energy exchange in the biosphere. Harper & Row, New York. 151 p.

————. 1963. The energy environment in which we live. Amer. Scientist 51:327–348.

————. 1964. Leaf temperature and transpiration. Agron. J. 56:273–277.

————. 1965. Energy, plants and ecology. Ecology 46:1–13.

————, R. Alderfer, and E. Taylor. 1968. Leaf temperatures of desert plants. Science 195:994–995.

Gibbs, J. G., and D. T. Patten. 1970. Heat flux and plant temperatures in a Sonoran Desert ecosystem. Oecologia 5:165–184.

Hadley, N. F. 1972. Desert species and adaptation. Amer. Sci. 60:338–347.

Halvorson, W. L., and D. T. Patten. 1974. Moisture stress in upper Sonoran Desert shrubs in relation to soil moisture and topography. Ecology 55:173–177.

Huber, B. 1956. Die Temperatur pflanzlicher Oberflachen. *In* Handbuch der Pflanzenphysiologie 3:285–292. Berline, Gö Hingen, Heidelberg.

Idso, S. B., and D. G. Baker. 1967. Relative importance of reradiation, convection and transpiration in heat transfer from plants. Plant physiol. 42:631–640.

————, and R. D. Jackson. 1969. Thermal radiation from the atmosphere. J. Geophys. Res. 74:5397–5403.

Lange, O. L. 1965. Leaf temperatures and methods of measurement. *In* Methodology of Plant Eco-Physiology: Proceedings of the Montpellier Symposium, pp. 203–209 UNESCO.

maciformis was made, based upon CO_2 exchange measurements in the field. The annual photosynthetic net gain, which is chiefly due to moistening by dew, allows for a biomass increase in the range of 5 to 10 percent. This figure is confirmed by growth measurements with these desert lichens in the field.

INTRODUCTION

Deserts are characterized by high temperatures and lack of moisture. Only plants that have developed special adaptations for these extreme conditions are able to live in desert habitats. Evolution has followed several entirely different courses in generating plants that meet the requirements of the desert environment. There are two main strategies for ecological adaptation of plants to drought and extreme temperatures (Evenari et al., 1974): (1) "Arido-active" plants are metabolically active with fully hydrated tissues even during the hot, dry season. They have special characteristics that reduce the effects of excessive water loss and overheating of their leaves. (2) "Arido-passive" plants limit their metabolic activity to moist and humid periods only. They become more or less dehydrated during the driest, hottest periods, which leads to an anabiotic state. Deciduous perennials belong to this group of desert plants and shed their leaves during the dry period. Annuals, another type of arido-passive plants, survive the dry season in the insensitive stage of a seed. They complete their active life cycle entirely during the rainy season. There are lower plants that are better adapted than higher plants to repeated changes between activation during favorable and inactivation during unfavorable conditions. Lichens play a dominant role within this group. Their exceptionally successful adaptations to drought and extreme temperatures mainly explain why lichens are found in deserts around the world and why they are important members of these different desert ecosystems.

What are the special physiological adaptations that enable the lichens to live under hostile desert conditions? We tried to answer this question by conducting investigations in the field, in the Negev Desert in Israel, and in the laboratory. Two main postulations must be fulfilled for a poikilohydric organism to exist in a hot, dry habitat. First, the plant must be able to maintain sufficiently positive photosynthetic productivity in spite of the extreme environment. Second, the organism's resistance to drought and high temperatures must be adequate to avoid injury from these factors. The task is to find out in what respect the lichens are adapted to meet these requirements.

The work summarized in this report is described and discussed in detail in earlier publications (Lange, 1965, 1969a, 1969b; Lange et al., 1969, 1970a, 1970b; Lange and Evenari, 1971; Kappen and Lange, 1972).

Information on methods used in these investigations may be obtained from papers by Lange et al. (1969), Koch et al. (1971), and Schulze et al. (1972). For a description of the habitat near Avdat/Negev Desert (Israel) where the field work was carried out, see Evenari et al. (1971).

We gratefully acknowledge the help of P. Rundel (Irvine, Calif.) and B. R. Strain (Durham, N.C.) who kindly read the manuscript of this paper.

PHOTOSYNTHESIS IN RELATION TO ENVIRONMENTAL CONDITIONS

Lichens lack roots and water transport systems. They are exclusively dependent on water uptake from their surroundings by the whole thallus surface. Such plants are called poikilohydric. Their internal water potential changes with the change of the ambient water potential. As a result of the meager precipitation on a desert site, the dehydrated thalli are moistened and metabolically activated by rainfall on only a few days of the year. This is scarcely sufficient to maintain adequate dry matter production. The question, therefore, is what other moisture sources are available to lichens in these arid regions, and in what manner are they assured of the necessary photosynthetic assimilation. The treatment of these problems was undertaken in extensive laboratory experiments, in which the CO_2 exchange of lichens was investigated in relation to environmental factors. The derived laboratory conclusions were examined and tested more closely under field conditions.

It is likely that dew condensation in the desert provides sufficient water to activate photosynthesis in lichens. To ascertain what degree of moistening is necessary for activation of the metabolic functions, net photosynthesis and dark respiration in relation to the moisture conditions of the lichen thallus was examined. Figure 1 shows the results of such an experiment with the desert species *Ramalina maciformis.* The lichen releases small quantities of CO_2 even in light at low water content. It reaches its moisture compensation point at a water content level of approximately 20 percent. Net photosynthesis increases with increasing water content until maximal rates are reached at about 60 percent. The arrows on the abscissa of Figure 1, which range from approximately 40 to 90 percent water content, are values of moistening measured in samples of the same lichen species after dewfall under natural conditions in the Negev Desert. These results show that this lichen may reach moisture levels after dew in the field that enable relatively high or maximum photosynthetic activity.

The question of whether this ability is actually realized in the natural habitat can only be solved by field measurements. Thus, after natural moistening as a result of dewfall, lichens were installed, shortly before sunrise, in plant chambers in which their CO_2 exchange could be fol-

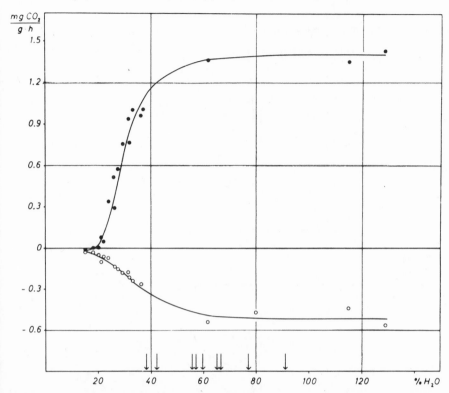

FIGURE 1   Laboratory measurements of CO_2 exchange of *Ramalina maciformis* during light (dots, 10,000 lx intensity) and dark respiration (circles) in relation to moisture state of the lichen (temperature 10°C). Abscissa: water content of thalli related to dry weight. Ordinate: quantities of carbon dioxide absorbed or released, respectively. The arrows along the abscissa indicate the water content of the lichen following dew wetting under natural conditions during various seasons at Avdat/Negev Desert. (After Lange, 1969a.)

lowed and examined with infrared gas analyzers. The gas-exchange chambers were made of Plexiglas. The temperature and humidity conditions inside each chamber were adjusted by a Peltier conditioning system to exactly match ambient conditions. Figure 2 illustrates one experiment on *Ramalina maciformis* in the Negev Desert. At 4 A.M. the dew-moistened lichens were placed in the chamber. The thalli respired at first until photosynthesis increased as a result of the increasing illumination. The compensation point was passed, and at 7 A.M. the CO_2 uptake reached its maximum rate. After this, the lichen became increasingly dehydrated as a result of increasing radiation and temperature. The CO_2 absorption decreased sharply until 8:30 A.M., when the compensation point once again was surpassed and the lichen began to

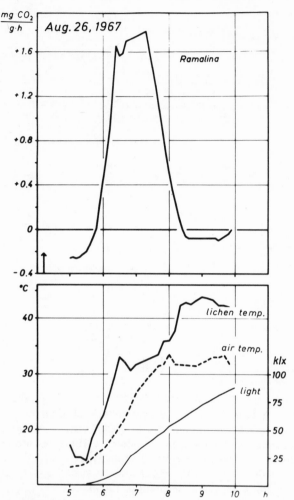

FIGURE 2 CO_2 exchange (above) of *Ramalina macifor-mis;* lichen temperature, air temperature, and light intensity (below) after moistening with dew. Upper ordinate: see Figure 1. Abscissa: time of day (A.M.). Arrow: time of introduction of lichen thalli into gas exchange chamber. Avdat/Negev Desert: Aug. 26, 1967. (After Lange et al., 1970a).

respire again. Respiration subsided, thereafter, because of increased desiccation.

The duration and intensity of photosynthesis after dew condensation is dependent on evaporation factors. There is some doubt whether the natural relationships are fully reached during the experiment shown, as the water loss of the dew-moistened lichens inside and outside the plant chamber might vary despite the fact that temperature and humidity are the same. We, therefore, performed a second series of experiments in which a number of lichen samples were examined for short periods of time. Several plant chambers were used to conduct short-term measurements with new samples being frequently exchanged for those previously used. Water content of the thalli was determined simultaneously with the photosynthesis measurements. The drying of the thalli enclosed in the chambers agreed well with the water loss of the specimens outside on the test site. The CO_2 exchange of successively measured lichen samples on a day with relatively little dew is represented separately in the curves of Figure 3. The integrated curve shows the behavior corresponding to natural conditions. It agrees well with the previous results.

These field studies basically confirm the laboratory conclusions. Nightly dewfall reactivates the lichens in their natural habitats and leads at least for a short time to high degrees of photosynthetic activity. This result does not only apply to the fruticose lichens of the *Ramalina* type. The same results were found for all 10 species of lichens studied in the Negev, including other lichen growth forms.

In addition to the moisture that results from nightly dewfall, a second water source, the high humidity level of the air during night, must be considered. We found that desiccated desert lichens are capable of obtaining enough water to become physiologically active solely from air with a high water vapor pressure. Reactivation of the lichens is possible even at water potentials much lower than saturated air. Gas exchange was measured in the laboratory in lichens whose water potential remained in equilibrium with air of a given relative humidity. One experiment with *Ramalina maciformis* is depicted in Figure 4. Respiration was measurable only when water potential equalled that of air of 75 percent relative humidity. At 80 percent relative humidity, which corresponds to a water potential of about -290 atm, however, the lichen reaches its moisture compensation point. Net photosynthesis at 90 percent relative humidity amounts to one fourth of the maximum attainable at 100 percent relative humidity. The fact that photosynthetic activity occurs after humidity uptake without any liquid water being available, and that productivity begins at a water potential as low as -290 atm illustrates the efficiency of lichen organization.

The speed and intensity of reactivation of lichen metabolism by humidity uptake are dependent on the special convection conditions during experiments in the laboratory. Again, determination of the ecological significance of these results is only possible through field research. Therefore, the CO_2 exchange of desert lichens was followed in their natural habitat in the morning after dew-free nights, but with the relative humidity still significantly high. One of these experiments is shown in Figure 5. The water content of the lichen in the early morning

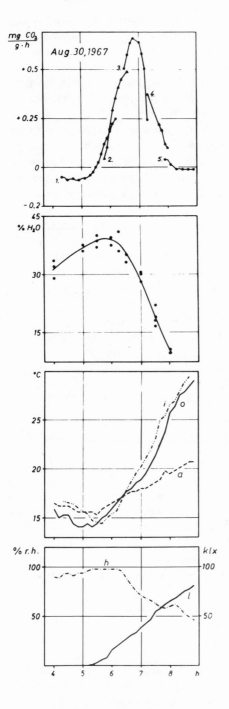

FIGURE 3 Top: CO_2 exchange of dew-moistened *Ramalina maciformis*. Second: water content of the thalli. Third: temperature of the air *(a)*, temperature of the lichens in gas-exchange chamber *(i)*, and outside at the test site *(o)*. Bottom: light intensity *(l)* and relative humidity *(h)* of the air. Numbered curve segments: responses of the different thalli after reloading the chambers with fresh lichen specimens from the test site. Abscissa: time of day (A.M.). Avdat/Negev Desert: Aug. 30, 1967. (After Lange et al., 1970a.)

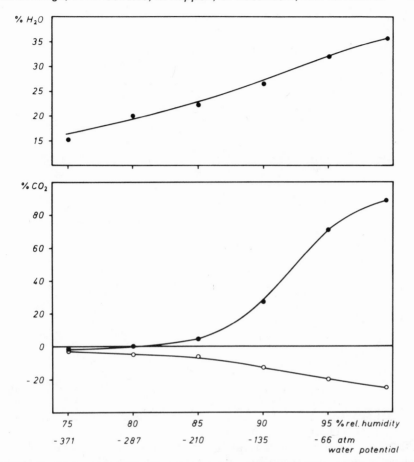

FIGURE 4 Water content (above) and CO_2 exchange (below) of *Ramalina maciformis* in equilibrium of water vapor pressure with air having different relative humidity and water potential, respectively (abscissa). Given are the quantities of carbon dioxide absorbed or released, respectively, during light (dots, 10,000 lx light intensity) and dark respiration (circles) in percentage of apparent CO_2 uptake following spraying with water. (After Lange, 1969.)

hours increased to a level of 31 percent solely as a result of water vapor uptake. That was sufficient for positive net photosynthesis for about 3 hours, after which time the thalli dried out and became inactive again.

RESISTANCE TO CLIMATIC EXTREMES

Periods of morning activation in the desert either by dew moistening or by water vapor uptake are terminated by high temperatures and strong desiccation brought about by extreme solar radiation. Desert

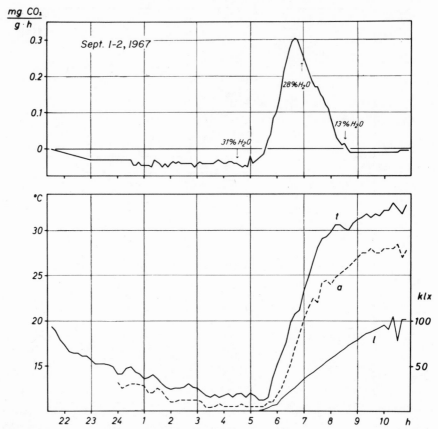

FIGURE 5 CO_2 exchange (above) of *Ramalina maciformis* in the course of a day with water vapor absorption without dew condensation; light intensity *(l)*, temperature of air *(a)*, and lichens *(t)* (below). Abscissa: time of day. Avdat/ Negev Desert: Sept. 1/2, 1967. (After Lange et al., 1970a.)

lichens have developed an unusual tolerance of these situations without suffering any damage. They are capable of surviving long periods almost completely dehydrated. This was shown by experiments in which lichen material was exposed to artificially reproduced dry periods, after which the thalli were rewetted and their photosynthetic activity measured. Figure 6 shows that even after about 1 year of severe desiccation *Ramalina maciformis* remains undamaged. After remoistening, the lichen again attains 100 percent of its original photosynthetic capacity. In any case, this resistance to desiccation allows the desert lichens to withstand the most extreme conditions of their natural habitat without injury or loss of time for reactivation. Recovery of the photosynthetic

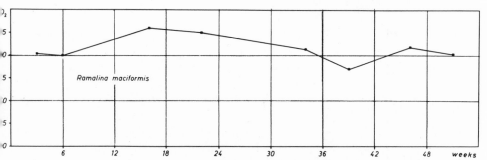

FIGURE 6 CO_2 exchange of *Ramalina maciformis* (10,000 lx light intensity, temperature 10°C) following drying periods (water content 1 percent). Abscissa: duration of the drying period. Ordinate: quantities of CO_2 uptake in percentage of apparent net photosynthesis before drying treatment. The figure shows maximal rates during the course of subsequent culturing. (After Lange, 1969.)

process after long periods of desiccation is possible solely by the absorption of water vapor.

In addition to their high drought resistance, desert lichens are extremely insensitive to heat while desiccated. The following results are from experiments in which strongly dehydrated lichen material was heated for ½ hour at a specific temperature. After this treatment the photosynthetic capability of the rewetted lichens was examined (Fig. 7). Treatment at temperatures up to 65°C had absolutely no effect on photosynthetic activity; following rewetting, activity was approximately 100 percent of the initial rate. Only with temperatures between 65 and 70°C did a noticeable depression occur. *Ramalina maciformis* was not killed until after exposure to 85°C and higher. Other species of lichens are able to withstand still higher temperatures. If we compare these values of heat resistance with the temperatures of lichen thalli in their natural habitat, we may conclude that there is no danger of heat injury.

High temperatures are not a constant condition at the desert site. On the contrary, the soil surface during winter may reach temperatures far below 0°C. Temperatures such as these are injurious for many higher plants. Experiments show, however, that desert lichens are affected neither in their vitality nor in their metabolic activity after treatment with very low temperatures. The frost resistance of *Ramalina maciformis,* for instance, shows practically no limits. In desiccated or even hydrated conditions this lichen can endure cooling for 6 hours at the temperature of liquid nitrogen without being affected in photosynthetic capability.

These findings clearly show how well the desert lichens are adapted to the extreme climatic conditions of their environment. Photosynthetic

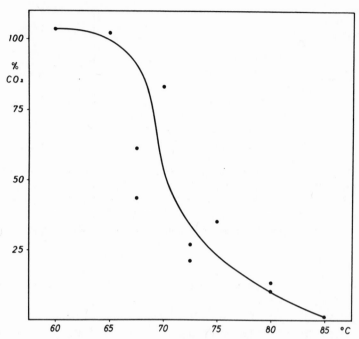

FIGURE 7 "Gross" photosynthetic CO_2 uptake of *Ramalina maciformis* following heating of the desiccated lichen for 30 minutes. Abscissa: temperature of
heating. Ordinate: CO_2 uptake in percentage of rate shown before heat treatment. (After Lange, 1969.)

activity can be almost immediately induced through moistening following long dry periods, after extreme heating periods under dry conditions, and after frost periods in winter. Most higher plants would be
killed or at least would suffer from marked inactivation of their
metabolic processes under similar circumstances.

The special adaptations of desert lichens to high temperatures, however, exist only under dehydrated conditions. Under moistened and
metabolically active conditions, lichens show a pronounced sensitivity
to higher temperatures and some adaptation to relatively low temperatures. This is expressed by the dependence of their CO_2 exchange on
temperature. Figure 8 shows laboratory measurements of net photosynthesis and respiration of *Ramalina maciformis* in relation to thallus
temperatures. Under illumination with 8800 lx, optimal CO_2 assimilation
occurs at a temperature of about 10°C. The peaks of the response
curves shift into the region of higher temperatures with increasing light
intensities. The optimal temperatures, however, do not exceed 20°C.
Such optimal range appears at first to be very low for an organism
whose natural habitat is a hot desert. However, periods of actual

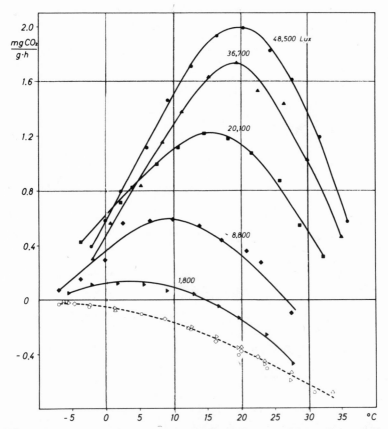

FIGURE 8 Temperature dependence of CO_2 exchange of *Ramalina maciformis* at different light intensities (dots) and dark respiration (circles). Abscissa: thallus temperature. (After Lange, 1969.)

metabolic activity for these lichens are restricted to the cool morning hours just following sunrise, after being moistened by dew or by high humidity. During this time, the thallus temperature does not reach far above 25°C in the summer, and during winter it seldom reaches 10°C for the same period of the day. This temperature range is optimal for the photosynthetic productivity of the desert lichen, especially under conditions of low solar radiation.

Adaptation of hydrated desert lichens to comparatively low temperatures is apparent, too, in their heat resistance. In contrast to dry conditions, moist lichens are very sensitive to increased temperatures. This is illustrated in Figure 9 by one set of experiments with *Ramalina maciformis*. The lichen material was warmed in the laboratory for 1 hour under completely hydrated conditions. After this treatment its photosynthetic activity under the routine experimental conditions was meas-

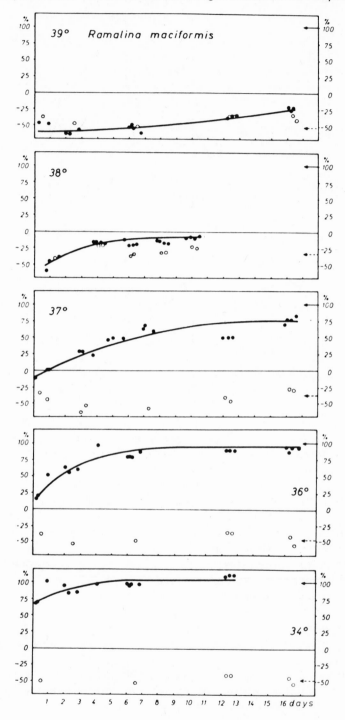

ured and followed for some time. At 34°C, photosynthetic capability remained practically unchanged at 100 percent of its original rate. Above 37°C, a remarkable depression was induced, which, at the end of a recovery period of 16 days, almost disappeared. Exposure to 38 or 39°C, however, heavily damaged or completely destroyed the photosynthetic mechanism. It is ironic that plants from hot desert environments when fully hydrated perish under such moderate temperatures. This fact illustrates dramatically the uniqueness of the poikilohydric character.

PHOTOSYNTHETIC PRODUCTIVITY

We have discussed how and under what conditions desert lichens are able to photosynthesize and how they are adapted to survive extreme climatic situations. However, the ultimate test for survival of the lichen in the desert is metabolic balance—the net production over long periods of time. To determine daily photosynthetic yield, experiments were carried out in the Negev Desert on *Ramalina maciformis*. CO_2 exchange and water relations were monitored during and after dewfall for periods of 24 or 36 hours. One example is given in Figure 10. During day-time no CO_2 exchange was detected because the lichens were fully dehydrated. Nocturnal respiration became measurable at 8 P.M. when the lichen began to absorb water vapor from the ambient air. The respiration continued throughout the night with remarkable intensity. In the early morning hours, after dewfall, the lichen had a maximum water content of about 60 percent. This enabled a 3-hour period of photosynthetic activity after sunrise. Subsequently, the compensation point was surpassed and the lichen again exhibited some respiration. The dry, inactive state was reached at noon and lasted until the humidity level during the evening hours again reactivated the respiratory process. This experiment shows very clearly how the lichens in the desert must live on their "morning breakfast" only.

The extent of the daily gain in net production after dewfall is largely dependent upon the length of the active period. In other words, it depends upon when light intensity causes the lichen to pass the compensation point and how soon desiccation takes place. Both factors are related to the lichen's exposure. To show this variability, two gas-

FIGURE 9 CO_2 exchange of *Ramalina maciformis* during light (dots, 10,000 lx light intensity) and dark respiration (circles) following heating of the moist lichen for 60 minutes; heating temperature indicated. Abscissa: time after heat treatment. Ordinate: quantities of CO_2 absorbed or released, respectively, in percentage of net photosynthesis shown before heat treatment. (After Lange, 1965.)

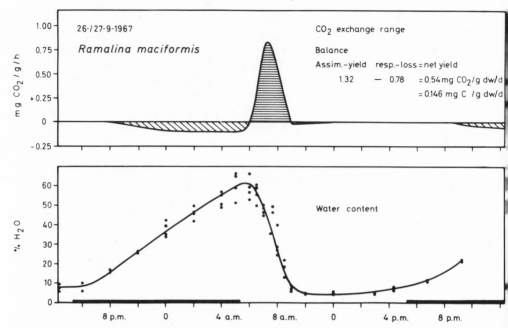

FIGURE 10 CO_2 exchange (average of several measurements) of *Ramalina maciformis* (above) in the course of a day with dew moistening; water content of the thalli (below). Abscissa: time of day. Avdat/Negev Desert: Sept. 26/27, 1967. (After Lange, 1969b.)

exchange chambers were exposed in different directions, one facing east and the other north. The photosynthetic response curves of the lichens in the two chambers after dew moistening were entirely different (Fig. 11). Light intensity was high before noon in the east exposure, but was low in the north exposure. This resulted in large differences in the lichen temperatures, which consequently implied different rates of desiccation. The lichen facing north photosynthesized many hours longer than did the east-facing lichen. In the latter, dehydration caused the CO_2 exchange to fall below the compensation point by 7:30 A.M. In spite of lower absolute rates of CO_2 uptake, the productive yield of the lichen facing north was almost twice as high as in the east-facing lichen. This advantage is apparently the reason why the lichen vegetation in the Negev Desert is especially prominent on slopes with a northern exposure.

Figure 11 shows the response of *Ramalina maciformis* after a comparatively high amount of dewfall, whereas Figure 10 represents the situation with an average dewfall for the central Negev Desert. This experiment might provide an estimation of the daily CO_2 balance of the lichen growing in a horizontal exposure. The assimilatory yield during that typical day amounts to 1.32 mg of $CO_2 \cdot g^{-1}$ (dry weight); the res-

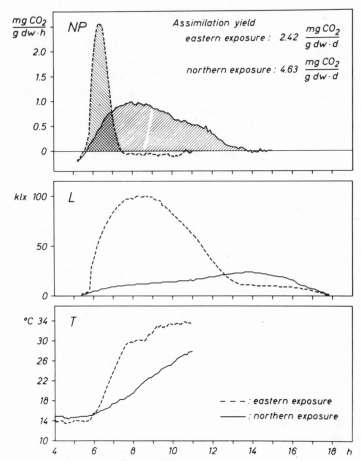

FIGURE 11 CO_2 exchange of *Ramalina maciformis (NP)* after moistening with dew; light intensity *(L);* thallus temperature *(T).* Thalli exposed to the north (solid line) and to the east (dashed line). Abscissa: time of day. Avdat/Negev Desert: Sept. 14, 1971.

piratory loss is 0.74 mg of $CO_2 \cdot g^{-1}$. The net yield, as computed from these values, amounts to 0.54 mg of $CO_2 \cdot g^{-1} \cdot d^{-1}$. Chemical analysis has shown that the dry thalli of *Ramalina maciformis* consist of 34 percent organic bound carbon. A net gain of 0.143 mg of carbon per gram of lichen during 1 day with an average dewfall therefore means a carbon fixation that would allow a biomass increase of 0.41 percent. This number allows a crude estimation of the annual productivity of the lichen. At the measuring site in the Negev Desert, there are on the average 198 nights per year with dewfall. If one assumes that every dew night results in the same net production as measured for the average situation in September, it would appear that a yearly biomass increase of 8.4 percent is possible, which is strictly due to the activation of the lichens by dewfall. In comparison, the CO_2 gain of the lichen in response to the few rainfalls in the Negev Desert can be neglected. The photosynthetic production that is possible through water vapor uptake in nights without dew condensation seems to be sufficient to compensate for the remaining respiratory loss of the plants. Thus, it follows from this general approximation that the total annual growth rate of *Ramalina maciformis* in the Negev Desert ranges in the order of magnitude of 5 to 10 percent. This figure, based on CO_2 exchange measurements, was confirmed by direct growth rate measurements with other lichen species at the same site over a period of 5 years. The biomass production of *Ramalina maciformis* is very remarkable for a plant under the extreme conditions of the desert climate. The high efficiency of lichens is the result of special adaptations of these poikilohydric, "arido-passive" organisms to drought and extreme temperatures at the desert site.

LITERATURE CITED

Evenari, M., O. L. Lange, L. Kappen, E.-D. Schulze, and U. Buschbom. 1975. Adaptive mechanisms in desert plants. In Proceedings of the AIBS-meeting 1973 (in press).

————, L. Shanan, and N. Tadmor. 1971. The Negev. The challenge of a desert. Harvard University Press, Cambridge, Mass. 345 p.

Kappen, L., and O. L. Lange. 1972. Die Kälteresistenz einiger Makrolichenen. Flora 161:1–29.

Koch, W., O. L. Lange, and E.-D. Schulze. 1971. Ecophysiological investigations on wild and cultivated plants in the Negev Desert. I. Methods: A mobile laboratory for measuring carbon dioxide and water vapour exchange. Oecologia 8:296–309.

Lange, O. L. 1965. Der CO_2 -Gaswechsel von Flechten nach Erwärmung im feuchten Zustand. Ber. Dtsch. Bot. Ges. 78:441–454.

————. 1969a. Experimentell-ökologische Untersuchungen an Flechten der Negev-Wüste. I. CO_2-Gaswechsel von *Ramalina maciformis* (DEL.) BORY unter kontrollierten Bedingungen im Laboratorium. Flora, Abt. B 158:324–359. [Technical translation 1654, National Research Council of Canada: Ecophysiological investigations on lichens of the Negev Desert. I. CO_2 gas exchange of *Ramalina maciformis* (DEL.) BORY under controlled conditions in the laboratory.]

————. 1969b. Die funktionellen Anpassungen der Flechten an die ökologischen Bedingungen arider Gebiete. Ber. Dtsch. Bot. Ges. 82:3–22.

————, and M. Evenari. 1971. Experimentell-ökologische Untersuchungen an Flechten der Negev-Wüste. IV. Wachstumsmessungen an *Caloplaca aurantia* (PERS.) HELLB. Flora 160:100–104.

———, W. Koch, and E.-D. Schulze. 1969. CO₂-Gaswechsel und Wasserhaushalt von Pflanzen in der Negev-Wüste am Ende der Trockenzeit. Ber. Dtsch. Bot. Ges. 82:39–61.

———, E.-D. Schulze, and W. Koch. 1970a. Experimentell-ökologische Untersuchungen an Flechten der Negev-Wüste. II. CO₂-Gaswechsel und Wasserhaushalt von *Ramalina maciformis* (DEL.) BORY am natürlichen Standort während der sommerlichen Trockenperiode. Flora 159:38–62. [Technical translation 1655, National Research Council of Canada: Ecophysiological investigations on lichens of the Negev Desert. II. CO₂ gas exchange and water relations of *Ramalina maciformis* (DEL.) BORY in its natural habitat during the summer dry period.]

———, E.-D. Schulze, and W. Koch. 1970b. Experimentell-ökologische Untersuchungen an Flechten der Negev-Wüste. III. CO₂-Gaswechsel und Wasserhaushalt von Krusten- und Blattflechten am natürlichen Standort während der sommerlichen Trockenperiode. Flora 159:525–538. (Technical translation 1656, National Research Council of Canada: Ecophysiological investigations on lichens of the Negev desert. III. CO₂ gas exchange and water relations of crustose and foliose lichens in their natural habitat during the summer dry period.)

Schulze, E.-D., O. L. Lange, and G. Lembke. 1972. A digital registration system for net photosynthesis and transpiration measurements in the field and an associated analysis of errors. Oecologia 10:151–166.

DYNAMICS OF GREAT BASIN SHRUB ROOT SYSTEMS

Martyn M. Caldwell and Osvaldo A. Fernandez

*Range Science Department and the Ecology
Center, Utah State University, Logan, Utah*

Abstract

Communities dominated by cool desert shrubs in northern Utah exhibit a particularly heavy commitment of carbon to the belowground plant system. This is reflected both in high root–shoot biomass ratios on the order of 9 and in annual reconstruction of a significant fraction of the lateral root system. Individual root elements of most of the lateral root system undergo meristematic activity and growth for usually only 2 weeks or less. Although individual root apical meristems may be limited in time of activity, the entire root system is active for most of the year as compared to a very short season of activity for shoot growth aboveground. There is a progression of increasing root growth activity with depth in the soil profile from March through the summer season and into the fall. Root element extension has been observed under extreme water potentials of -70 atm and also in the middle of winter with soil temperatures on the order of 2 to $5°C$ when the uppermost layers of the soil were solidly frozen. The carbon costs of root system maintenance and growth activity are considerable when compared to aboveground growth activity.

INTRODUCTION

Although the dynamics of plant root systems in nonagricultural communities have received generally inadequate study, the case of Great Basin shrub-dominated communities draws particular attention since approximately 90 percent of the shrub biomass is belowground. The relative importance of the belowground system for these cool desert, shrub-dominated communities becomes apparent in a simple comparison of plant standing carbon pools with those of a mesic hardwood-forest ecosystem in Tennessee (Reichle et al., 1973). For example, whereas the *Atriplex*-dominated community possesses only 2 percent of the aboveground plant carbon pool of that of the hardwood forest, it maintains 78 percent of the belowground plant carbon pool (see Fig. 1). Since the estimated annual carbon fixation of this *Atriplex* community is only about 7 percent of that of the hardwood forest, the relative carbon costs for maintenance of this sytem are proportionately quite high. In addition to maintenance respiration costs of these root systems, carbon investment in annual reconstruction of a portion of the root system is also now appearing as a significant component of our preliminary carbon balance studies of these shrub-dominated systems.

In this paper we summarize the salient aspects of our belowground plant studies for communities dominated by *Atriplex confertifolia* (Torr. and Frem.) S. Wats, *Ceratoides lanata* Nevski, and *Artemisia tridentata* Nutt. ssp. *wyomingensis* Beetle in Curlew Valley of northern Utah (113°5'W, 41°5'N, elevation 1350 m). Assuming that at least some element of energy cost efficiency has been of selective value in the evolu-

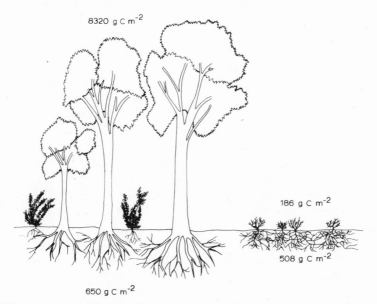

$8320 \ g \ C \ m^{-2}$

$186 \ g \ C \ m^{-2}$

$508 \ g \ C \ m^{-2}$

$650 \ g \ C \ m^{-2}$

FIGURE 1 Comparative standing carbon pools in the above- and belowground biomass of a mesic hardwood forest ecosystem at the Oak Ridge National Laboratory (Reichle et al., 1973) and a cool desert ecosystem dominated by *Atriplex confertifolia* in 1973.

tion of these species, it might be hypothesized that the unquestionably heavy commitment of carbon to the belowground plant system is quite necessary for survival in this harsh environment.

THE COOL DESERT ENVIRONMENT

Cool desert shrubs exist in mixed or sometimes nearly monospecific stands with respect to the perennial vegetation that comprises most of the plant biomass in these communities. The mean annual precipitation at the study site is approximately 230 mm, which may seem somewhat luxuriant for a true desert. Whether these communities are considered as simply arid shrub steppes or cool desert is a matter of interpretation. However, a definite element of aridity is superimposed in the form of high concentrations of salt, which are particularly pronounced in the *Atriplex*- and *Ceratoides*-dominated areas. *Artemisia* is usually not considered halophytic and is only occasionally found in soils of appreciable salinity. The soils in the basin of Curlew Valley are lacustrine in origin, having been formed about 7000 years ago in our study sites (Eardley et al., 1957) by the recession of Lake Bonneville, the remnant of which is now the Great Salt Lake. During the driest portions of the year, salt concentrations in the soil solution can comprise as much as -35 atm of the total soil water potential (Moore and Caldwell, 1972). Air temperatures during summer months often range as high as 40°C, which is not appreciably cooler than many warm desert areas. However, during the cooler remainder of the year, temperatures are often below freezing and frequently reach extremes of -30°C. Although the soil environment is more moderate in terms of both the extremes and fluctuations of water potential and temperature as compared to the aerial plant environment, it does constitute a very severe milieu for root growth and effective absorption of moisture and nutrients. The seasonal courses of soil moisture and soluble salt concentrations are depicted in Figure 2. Soil moisture recharge results primarily from winter precipitation in the form of snow. Summer season precipitation is of significance only for recharge of the uppermost soil layers. Therefore, during the course of the primary growing season from March through September, there is a steady decline of soil moisture in most of the profile. Soil moisture potentials during the drier portion of the year of less than -70 atm are quite common in these soils (Moore and Caldwell, 1972; Fernandez and Caldwell, 1975b).

ROOT AND SHOOT BIOMASS

The *Atriplex confertifolia* and *Ceratoides lanata* dominated communities have exhibited very high root–shoot ratios (Bjerregaard, 1971; Fernandez, 1974). In July 1973, for example, 87 percent of the *Atriplex*-

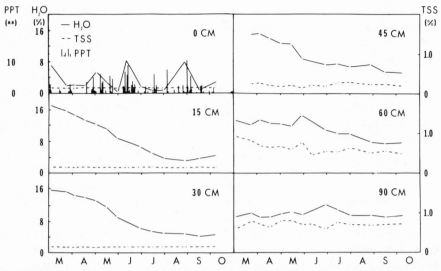

FIGURE 2   Soil moisture content and total soluble salts (TSS) at several depths in a community dominated by *Atriplex confertifolia* and *Ceratoides lanata* in Curlew Valley. Precipitation during the same period is shown on the graph for the soil surface level (0 cm). Total soluble salt concentrations are represented from a standard soil paste used in a Beckman conductivity bridge. (From Moore and Caldwell, 1972.)

dominated community plant biomass was belowground and 92 percent of that of the *Ceratoides*-dominated community. This corresponds to root–shoot ratios of 6.7 and 11, respectively (Fernandez and Caldwell, unpublished data). In a recent review article, Barbour (1973) questioned the often-iterated tenant that desert plants, or xerophytes in general, exhibit high root–shoot biomass ratios. He presented data from several authors indicating that in fact root–shoot ratios of many warm desert plants were substantially less than unity. In an extensive review of plant biomass determinations from desert and steppe areas, primarily in the USSR, Rodin and Bazilevich (1967) reported generally higher root–shoot ratios than those of the authors cited by Barbour (1973) from studies carried out in North America and the Middle East. For a variety of arid grass and shrub-dominated steppes in the USSR, Rodin and Bazilevich (1967) reported root–shoot biomass ratios comparable with those we have found for *Atriplex* and *Ceratoides.* This is particularly the case for saline steppe regions in the USSR.

METHODOLOGY OF ROOT BIOMASS DETERMINATION

Although the seemingly pedestrian techniques for extracting roots from soil receive little detailed attention in the literature, it has become apparent in our studies that very fine root elements are particularly difficult to extract and do indeed constitute a sizable component of total belowground biomass. For example, a series of tests comparing a flotation technique, the last step of which consisted of pouring suspended root segments through a sieve with a pore size of 0.2 mm^2, with a similar technique that involved flotation in saturated NaCl solution and pouring the suspended root material through a sieve with a pore size of 0.03 mm^2 indicated that the second technique yielded a higher biomass value by a factor of approximately 2.4. This was primarily due to the component of extremely fine root elements. Microscopic observations of these root material collections indicated that most of the material was indeed root tissues (Fernandez and Camp, unpublished data). Therefore, it is apparent that much more attention is warranted in considerations of techniques of extractions of roots from soils, particularly when very fine and profuse root systems are involved.

With the exception of large roots, the detection of living and dead roots on a large scale presents a difficult and disconcerting problem in field studies of plant root systems. Although some improvisations of techniques have been recently proposed, such as the use of triphenyl-tetrazolium chloride staining on a large scale (Knievel, 1973) or carbon 14 autoradiography (Singh and Coleman, 1973), each of these procedures requires rather extensive calibration techniques and certain disquieting assumptions for each plant species under investigation. Since most reports of belowground biomass in the literature do not discuss the problem of living and dead root separations, it must be presumed that all such reports of root biomass include both living and dead root components, particularly for the fine root system. In our own studies, summarized in a foregoing section, estimates of community belowground biomass include both living and dead roots. Preliminary attempts in our studies to separate living and dead root system components by differential density in methanol have recently given encouraging results when compared to carbon 14 techniques involving root tissue combustion and counting or autoradiography of individual roots. Furthermore, the method appears to be promising for large-scale field sampling studies. Preliminary results from the *Atriplex* and *Ceratoides* communities suggest that in midsummer when most of the belowground biomass values were determined for 1973, approximately 17 and 30 percent of these root systems, respectively, appeared to be composed of dead root elements. This technique certainly deserves further investigation and refinement before extensive employment in field studies as no two species appear to respond in exactly the same manner (Fernandez, 1974). If it is tentatively concluded that about 70 to 85 percent of the belowground biomass of these shrubs are indeed composed of living root elements, the energy costs both in terms of maintenance respiration and in annual reconstruction of a portion of the root system would be decidedly high, as will be discussed later.

ROOT GROWTH AND MORPHOLOGY

Root observation chambers have been employed in the field in nearly monospecific stands of *Atriplex confertifolia, Ceratoides lanata,* and *Artemisia tridentata* over the last 2 years. These chambers with inclined Plexiglas observation windows have permitted a nondestructive record of growth of individual roots as well as microscopic observations. Soil water potentials and temperatures immediately proximate to the Plexiglas windows have not differed significantly from those in the adjacent community (Fernandez and Caldwell, 1975b). The root morphologies of these species differ as much as the aboveground plant parts. However, there are characteristics common to all three species. Except for the main extension roots, individual root apices appear to be growing for usually no more than 2 weeks. Once activity of a particular root growing tip ceases, based on our observations to date, it does not again resume growth. However, development of lateral roots from the same root may well occur at a later time. The degree of branching, color, diameter, and degree of elongation of individual root elements varies considerably among the three species; however, individual growing elements of all three species seem to undergo a similar pattern of elongation and development of a zone of dense root hairs somewhat behind the active meristematic regions, followed by apparent suberization within approximately 7 to 14 days. Despite the darkening of color, indicating apparent suberization, root hairs are persistent in all three species and remain in *Atriplex* for more than 1 year (Fernandez and Caldwell, 1975a).

Persistence of root hairs for long periods of time in a variety of woody and herbaceous species has been previously reported by other workers (Jeffrey and Torrey, 1921; McDougall, 1921; Whitaker, 1923; Dittmer, 1938). The key question, of course, is whether these root hair elements are functional in water and nutrient absorption. Cowling (1969) carried out ^{36}Cl uptake studies on roots of *Atriplex vesicaria* and concluded that uptake of this isotope was primarily dependent on new root growth and indicated that older, apparently suberized portions of the root system were much less effective in nutrient absorption. Presumably persistent root hairs on suberized roots are of limited utility for absorption, although this remains an open question worthy of further attention.

Although in most cases individual root apical meristems apparently are seldom active for more than 2 weeks, the root system as a whole maintains a long period of active growth throughout the year. The general pattern of root growth activity for *Atriplex confertifolia* during 1973 is presented in Figure 3. At each 10-cm depth increment in the soil profile, the period of appreciable root growth activity is denoted by a horizontal line, with the relative growth rate of the root system indicated by the blackened curve. Relative growth rate in this case is denoted as

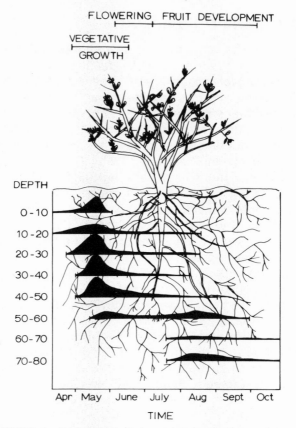

FIGURE 3 Extent and timing of root growth activity at several depths for a community dominated by *Atriplex confertifolia.* Horizontal bars for each depth increment represent the period during which appreciable root growth has taken place. Blackened curves represent relative growth rate of roots and are plotted as unit length of root growth per unit length of root mass apparent in the observation windows per day. The primary progression of phenological events of the aboveground portion of the plant are also represented. (Adapted from Fernandez and Caldwell, 1975a.)

unit length of growth per length of existing roots in particular sampling plots of the observation windows per day. This is calculated in a fashion analogous to relative growth rates of aerial plant parts as described by Kvet et al. (1971). In this manner, the growth rate measures are not biased by the amount of the root system directly apparent in the root observation chamber window.

Vegetative growth and phenological progression of the aboveground portion of the plant are also indicated in Figure 3. The major growth activity of the root system varies throughout the year depending on depth. This is apparently linked with the progression of soil moisture depletion and soil temperatures at various depths in the profile throughout the year. As mentioned earlier, primary soil moisture recharge is the result of winter precipitation, and summer rain only influences the uppermost layer of the soil profile. The depletion of soil moisture at various depths in the profile (see Fig. 2) corresponds well with regions of maximal root growth activity for various times during the season (Fig. 3). Roughly the same pattern of root growth activity during the seasons at various soil horizons was exhibited by all three shrub species. Similarly, aboveground growth and phenological progression of the three shrubs were also quite similar.

Although vegetative shoot growth was primarily active for only a short period during the season, root growth persisted for a much longer period during the year (Fig. 3). In fact, measurements taken as late as January 1974 indicated root growth activity when the upper soil horizons were solidly frozen and soil temperatures at greater depths were near freezing (Fernandez and Caldwell, unpublished data). Low soil temperatures on the order of 2 to 5°C at greater depths in the profile appear to limit root growth activity in the spring despite the presence of suitable soil moisture.

In addition to depleted soil moisture, soil temperatures exceeding 20°C may be limiting for root growth activity in the upper soil layers in mid to late summer (Moore and Caldwell, 1972; Fernandez and Caldwell, 1975a). During 1972, when soil moisture potentials were less than −70 atm at 40-cm depth in the *Atriplex*-dominated community, some root extension was still evident during these very dry periods (Fernandez and Caldwell, 1975b). The importance of the extension and formation of new roots for soil moisture acquisition has been suggested in many situations where soils are only partially wetted (Kramer, 1969; Slatyer, 1967). In this cool desert environment, the rather rapid apparent suberization of the fine root elements following initial extension, combined with the greatly reduced hydraulic conductivity of these soils of very low water potential (Slatyer, 1967), are likely compelling reasons for the necessity of continual root element formation and extension, and periodic reexploration of the same soil mass. The fact that *Atriplex confertifolia* appears capable of maintaining a greater root growth activity in the lower soil horizons during the driest portions of the year when soil moisture potentials may be −70 atm or less, may be one factor permitting continued positive net photosynthesis in August and September, as compared to a species such as *Ceratoides lanata,* which appears to be rather inactive in carbon dioxide gas exchange at this time of year (Caldwell, 1972). Under laboratory conditions with well-established potted plants, transpiration and positive net photosynthesis have been documented for both *Atriplex* and *Ceratoides* when soil moisture potentials were less than −70 atm (Moore et al., 1972b; White et al., unpublished data). Although both species have an inherent capability for water absorption at very low soil moisture potentials in the laboratory, under field conditions the differences in root growth activity, which are controlled by other factors as well as soil water potential, are probably a decisive factor in whether or not gas-exchange activity persists during drier periods. Other factors, however, such as the possession of C_4 photosynthesis and higher photosynthesis–transpiration ratios by *Atriplex* (Caldwell, 1972) or the proclivity of *Atriplex* to accumulate NaCl as opposed to K_2SO_4 of *Ceratoides* (Moore et al., 1972a), might also be factors contributing to the apparent differences in CO_2 exchange behavior of these two species during dry portions of the year.

Although continued extension of part of the root system during even the dry portions of the year can be easily hypothesized to be of adaptive significance for these cool desert shrubs, slow but continued root extension in mid-winter when the upper soil layers, and presumably the plant crown, are solidly frozen seems less explicable in an adaptive context.

CARBON COSTS OF ROOT SYSTEM MAINTENANCE AND ACTIVITY

Energy costs of root systems that are in steady state can be considered as consisting of two primary components: (1) maintenance respiration of all living root elements, and (2) energy costs associated with continual reconstruction of a portion of the root system, i.e., root system turnover. In situ root respiration measurements in the field cannot be accomplished directly because of the unresolved problem of segregating true root respiration from respiration of rhizosphere microorganisms. Even less feasible is the separation of maintenance respiration from respiration of roots associated with growth. Under laboratory conditions, potential root respiration rates can be obtained in short-term experiments even when roots are well established in unsterilized soil.

An example of the complete short-term carbon 14 budget of an *Artemisia tridentata* plant is illustrated in Figure 4. The plant was exposed to $^{14}CO_2$ for a 4- to 5-hour period on the first day. Partitioning of the photoassimilated carbon 14 in this closed-system experiment over an 18-day period was carried out by measuring $^{14}CO_2$ efflux from above- and belowground portions of the plant system and by making a complete carbon 14 inventory of the plant tissues and soil at the end of this

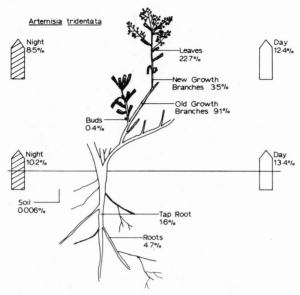

FIGURE 4 Carbon 14 allocation during an 18-day, closed-system, carbon 14 budget study of an individual *Artemisia tridentata* shrub. The plant was exposed to $^{14}CO_2$ for a 4- to 5-hour period on the first day. Percentage of carbon 14 is represented either as total carbon efflux during the night and day periods from the soil and aboveground system or as percentage of carbon recovered in combustion assay of plant tissues and the residual soil carbon.

period (Caldwell and Camp, unpublished data). Summation of carbon 14 collected from the gas efflux, plant tissues, and residual soil carbon accounted for 87 percent of the total photoassimilated carbon. The unaccounted for carbon 14 undoubtedly is the result of some measurement errors, small gaseous leaks, particularly from the belowground system, and incomplete harvesting of all carbon from the soil–root system.

Percentages of the total carbon 14 assimilated are represented in Figure 4. Although a small proportion of the total carbon was recovered from root system tissues, root respiration of $^{14}CO_2$ amounted to 23 percent of the carbon 14 assimilated at the beginning of the closed-system experiment. Because of the short-term nature of this experiment and the very small amount of carbon 14 left in the soil once roots were extracted, it is probable that most $^{14}CO_2$ efflux from the soil surface was indeed due to root respiration. Since the amount of carbon 14 in root tissues was comparatively small, it is possible that most respired $^{14}CO_2$ was due to maintenance respiration rather than synthesis respiration associated with growth. Assuming that the total percentage of carbon 14 recovered from root tissues was incorporated into new root growth, as opposed to temporary carbon storage, application of the theoretical and experimental "carbon dioxide production factors" of Penning de Vries (1972), which are on the order of 12 to 13 percent pure carbon respired as CO_2 per dry weight of plant material produced, indicates that approximately 2 percent of the total carbon 14 photoassimilated on the first day would have been expended in growth synthesis processes apart from maintenance respiration. Approximately 21 percent of the total carbon 14 assimilated on the first day would then have been expended on maintenance respiration during the subsequent 18 days.

If this pattern of carbon 14 allocation is representative of average total carbon partitioning, this rate might be indicative of a potential maintenance respiration rate. However, maintenance respiration under field conditions for these communities of very large belowground biomass must be considerably less than this potential rate or the whole-plant values (1.5 percent of plant dry matter per day) of maintenance respiration that appear in the literature (McCree, 1970; Penning de Vries, 1972).

Turnover of a certain fraction of the belowground plant system can also represent a sizable annual investment of carbon in the root system. Caldwell and Camp (1974) recently determined the fractional annual turnover rate of the belowground plant system for shrub communities dominated by *Atriplex confertifolia* and *Ceratoides lanata* for 1973. Determination of this annual turnover rate, termed a turnover coefficient, is based upon the change in the $^{14}C - ^{12}C$ ratio in cellulose tissues of the root system sampled on a community basis at the beginning and end of the primary season of root growth activity. This turnover coefficient is considered to be a conservative estimate of the total fraction of the root system that is reconstructed between the two sampling times. This has been carried out separately for the taproot and lateral root system. Although the *Ceratoides*-dominated community exhibited higher root–shoot ratios, the turnover coefficient of this community was only on the order of 12 percent, whereas it was on the order of 24 percent for the *Atriplex*-dominated community. The taproot elements in both communities exhibited much lower turnover coefficients, as would be expected for these generally longer lived elements of the root system. When turnover coefficients for the lateral diffuse root system are applied to the proportionately high root biomass estimates for these communities, it is apparent that the belowground turnover of productivity is of considerable magnitude—approximately three times that of the aboveground annual productivity. However, this still represents a conservative estimate of belowground productivity.

CONCLUDING REMARKS

We return briefly to a consideration of the rather high root–shoot biomass ratios of these cool desert communities as compared to several warm desert systems and the hypothesis that maintenance of this large proportion of underground root system is necessary for survival in the cool desert. Why should warm desert plant species be able to survive with root–shoot ratios less than unity (Barbour, 1973), whereas cool desert, shrub-dominated systems require proportionately much greater root biomass (root–shoot ratios on the order of 9) (Rodin and Bazilevich, 1967)? Certainly, overall annual water stress as determined by the balance between precipitation and evapotranspiration potential should be much greater in most warm desert areas. However, in a warm desert, many plant species undergo a drought-induced dormancy often manifested as drought deciduousness, whereas this is not so typical in cool desert shrub systems. Because low temperatures limit productivity during so much of the year in these cool desert systems, these plants simply may not be able to afford the luxury of becoming physiologically inactive during periods when water stress is most acute. Instead they would need to continue to absorb water wherever in the profile it might be most available. This in turn might depend on a large and very diffuse root system that maintains root growth activity during most of the year when temperatures are warm enough to permit physiological activity of the shoot system. Unfortunately, data on root turnover coefficients for other communities are not available for comparison. Such data would facilitate evaluation of overall carbon costs of maintaining root systems in warm and cool desert areas.

Certainly, our general knowledge of the dynamics, and even the living biomass, of desert root systems is quite crude and barely beyond the descriptive stage. Study of plant root systems in arid environments should hopefully undergo intensification both in quantitative refinement and in assuming a more functional approach at the whole-plant and community level.

Acknowledgments

This research was partly supported through the US/IBP Desert Biome Program (NSF Grant GB-32139), the Utah Agricultural Experiment Station, and a Ford Foundation Fellowship to O. A. Fernandez. The technical support of L. B. Camp is also gratefully acknowledged.

LITERATURE CITED

Barbour, M. G. 1973. Desert dogma reexamined: root/shoot productivity and plant spacing. Amer. Midl. Nat. 89:41–57.

Bjerregaard, R. S. 1971. The nitrogen budget of two salt desert shrub plant communities of western Utah. Ph.D. Dissertation. Utah State University, Logan, Utah. 100 pp.

Caldwell, M. M. 1972. Adaptability and productivity of species possessing C_3 and C_4 photosynthesis in a cool desert environment. In L. E. Rodin (ed.), Ecophysiological foundation of ecosystems productivity in arid zone, pp. 27–29. USSR Academy of Science, Leningrad, USSR.

————, and L. B. Camp, 1974. Belowground productivity of two cool desert communities. Oecologia 17:123–130.

Cowling. S. W. 1969. A study of vegetation activity patterns in a semiarid environment. Ph.D. Dissertation. University of New England, N.S.W. Australia. 286 p.

Dittmer, H. J. 1938. A quantitative study of the subterranean members of three field grasses. Amer. J. Bot. 25:654–657.

Eardley, A. J., V. Gvosdetsky, and R. E. Marsell. 1957. Hydrology of Lake Bonneville and sediments and soils of its basin. Bull. Geol. Soc. Amer. 68:1141–1202.

Fernandez, O. A. 1974. The dynamics of root growth and the partitioning of photosynthates in cool desert shrubs. Ph.D. Dissertation. Utah State University, Logan, Utah. 121 pp.

————, and M. M. Caldwell, 1975a. Phenology and dynamics of root growth of three cool semi-desert shrubs under field conditions. J. Ecol. 63 (in press).

————, and M. M. Caldwell. 1975b. Root growth of *Atriplex confertifolia* under field conditions. In J. Marshall (ed.), The belowground ecosystem: a synthesis of plant-associated processes. I. B. P. Synthesis Volume (in press).

Jeffrey, E. C., and R. E. Torrey. 1921. Physiological and morphological correlations in herbaceous angiosperms. Bot. Gaz. 71:1–31.

Knievel, D. P. 1973. Procedure for estimating ratio of live to dead root dry matter in root core samples. Crop Sci. 13:124–126.

Kramer, P. J. 1969. Plant and soil water relationships. McGraw-Hill, New York. 482 pp.

Kvet, J., J. P. Ondok, J. Necas, and P. G. Jarvis. 1971. Methods of growth analysis. In Z. Sestak, J. Catsky, and P. G. Jarvis (eds.), Plant Photosynthetic Production: Manual of Methods, pp. 343–391. Dr. W. Junk N. V. Publishers, The Hague.

McCree, K. 1970. An equation for the rate of respiration of white clover plants grown under controlled conditions. In I. Setlik (ed.), Prediction and measurement of photosynthetic productivity, pp. 221–230. Proc. IBP/PP tech. meeting, Trebon, 1969. Cent. Agric. Pub. and Doc., Wageningen, The Netherlands.

McDougall, W. B. 1921. Thick-walled root hairs of *Gleditsia* and related genera. Amer. J. Bot. 8:171–175.

Moore, R. T., and M. M. Caldwell. 1972. Field use of thermocouple psychometers in desert soils. In R. W. Brown and B. P. van Haveren (eds.), Psychometry in water relations research, pp. 165–169. Utah Agr. Exp. Sta.

————, S. W. Breckle, and M. M. Caldwell. 1972a. Mineral ion composition and osmotic relations of *Atriplex confertifolia* and *Eurotia lanata*. Oecologia 11:67–78.

————, R. S. White, and M. M. Caldwell. 1972b. Transpiration of *Atriplex confertifolia* and *Eurotia lanata* in relation to soil, plant, and atmospheric moisture stresses. Can. J. Bot. 50:2411–2418.

Penning de Vries, F. W. T. 1972. Respiration and growth. In A. R. Rees, K. E. Cockshull, D. W. Hand, and R. G. Hurd (eds.), Crop processes in controlled environments, pp. 327–347. Academic Press, New York.

Reichle, D. E., B. E. Dinger, N. T. Edwards, W. F. Harris, and P. Sollins. 1973. Carbon flow and storage in a forest ecosystem. In G. M. Woodwell and E. V. Pecan (eds.), Carbon and the biosphere. pp. 345–365. U.S. Atomic Energy Commission.

Rodin, L. E., and N. I. Bazilevich. 1967. Production and mineral cycling in terrestrial vegetation. Oliver & Boyd, Edinburgh. 288 p.

Singh, J. S., and D. C. Coleman. 1973. A technique for evaluating functional root biomass in grassland ecosystems. Can. J. Bot. 51:1867–70.

Slatyer, R. O. 1967. Plant–water relationships. Academic Press, New York. 347 p.

Whitaker, E. S. 1923. Root hairs and secondary thickening in the Compositae. Bot. Gaz. 76:30–59.

ENVIRONMENTAL AND PLANT FACTORS INFLUENCING TRANSPIRATION OF DESERT PLANTS

W. L. Ehrler

U.S. Water Conservation Laboratory, Phoenix, Arizona

Abstract

Experiments were carried out in a controlled environment chamber to measure the effects on transpiration and calculated leaf diffusion resistance (R_L) of illumination and saturation deficit (SD) for mesquite and of soil matric potential (ψ_M) for mesquite and jojoba. This work was supplemented by a greenhouse experiment with corn and agave to measure diurnal changes in transpiration, leaf-temperature differences, and R_L when ψ_M was high (soil was well-watered).

Data from the chamber were used in the transpiration equation to integrate the effects of both environmental and plant factors; the equation permitted calculation of R_L values from 1.8 to 87 s cm^{-1}, depending on species, illuminance, SD, and ψ_M. In an environment conducive to a high transpiration rate [illuminance, 44 kilolux (klx); air and leaf temperatures, 30°C; SD, 27 mb; ψ_M, −4 cb; and boundary layer resistance, 0.9 s cm^{-1}, transpiration rates of mesquite and jojoba were 230 g m^{-2} h^{-1} and 66 g m^{-2} h^{-1}; R_L values were 2 and 8.4 s cm^{-1}, respectively. Despite these distinct differences between species when ψ_M was high (not explainable by the present data), when ψ_M was low, both mesquite and jojoba developed sufficiently

This paper is a contribution from the Agricultural Research Service, U.S. Department of Agriculture.

high R_L values to limit transpiration severely, as expected for desert plants.

Corn responded in the greenhouse like a typical mesophyte with a high transpiration rate (an average of 90 with a peak at 175 g m^{-2} h^{-1}) and a relatively low R_L (7.1 s cm^{-1}) in the light, as contrasted to a low rate (6 g m^{-2} h^{-1}) and a high R_L (93 s cm^{-1}) in the dark.

However, even when ψ_M was high, the transpiration rate of agave was only 15 g m^{-2} h^{-1} in the light but doubled in the dark, consistent with respective R_L values of 239 and 17 s cm^{-1}. Presumably, agave functions in this manner because of its crassulacean acid metabolism, sunken stomates, and high impermeable cuticle.

INTRODUCTION

Transpiration is affected by many environmental factors (solar radiation, wind, water vapor pressure, and barometric pressure) and plant factors (cuticular permeability, stomatal responses, and type of metabolism). Solar radiation has two additive effects. It heats the leaf, thereby raising the internal vapor pressure, and opens the stomates, thus reducing the leaf diffusion resistance (R_L) to the escape of water vapor from the intercellular spaces to the air. However, under proper conditions (leaf temperature, T_L, exceeding air temperature, T_A) an increase in wind speed can cool leaves and more than compensate for the decrease in boundary layer resistance (R_A), and thus lower transpiration rate (Gates, 1968).

Despite the complex interrelations of environmental factors, their effect on transpiration can be made intelligible by using the transpiration equation (Ehrler and Van Bavel, 1968)

$$T = \frac{\Delta d_v}{R_A + R_L} \tag{1}$$

where T is the transpiration rate (g cm^{-2} s^{-1}), Δd_v the difference in water vapor density between the leaf interior and the air (g cm^{-3}), and R_A and R_L (s cm^{-1}) are as defined previously. This report presents data for the effect on the transpiration rate of (1) a given environmental factor that is varied while other pertinent factors are held constant, and (2) plant species growing in a wet soil and exposed to high solar radiation. The species tested have a broad range of transpiration rates: mesquite (very high), corn (high), jojoba (moderately low), and agave (extremely low). The hypothesis was that by understanding the differences among species in response to water stress the mechanism of stomatal control of transpiration might be discovered.

METHODS

Plant species. To obtain a wide range of transpiration rates a mesophyte, corn (*Zea mays* L.), and the xerophytes Chilean mesquite (*Prosopis chilensis* Stuntz), jojoba [*Simmondsia chinensis* (Link) Schneider], and century plant (*Agave americana* L.) were used. The corn was Mexican June, a cultivar highly adapted to the local climate, as are mesquite and agave, which were introduced from Chile and Mexico, respectively. Jojoba is a native of the Sonoran Desert of the southwestern United States and Mexico. The agave was grown from offshoots and the other species from seeds.

The test consisted of several short-term experiments with mesquite and jojoba in a controlled environment plant chamber, and a greenhouse experiment comparing corn with agave.

Plant Chamber. The chamber (22 cm × 61 cm × 61 cm) was built to accommodate large plants, and was located in the center of a large controlled environment room providing precise control of T_A (within 0.25°C), vapor pressure (within 0.1 mb), and illumination (within 2 percent). The top, bottom, and two sides of the chamber were made of 0.1-mm-thick polyvinyl chloride. The door (third side) was made of 6.6-mm-thick acrylic plastic and was held in place magnetically. The fourth side of the chamber was wooden with openings to accommodate a squirrel-cage blower and heat-exchanger coils for rapid circulation of cool air within the chamber.

Air temperature. In the dark the controlled environment room and the T_A of the plant chamber were in equilibrium. When the chamber lights were on, chilled water was pumped through the heat exchanger fast enough to absorb the radiant energy from a bank of mercury vapor lamps 1.5 m above the top of the chamber. The flow of chilled water was varied by manually adjusting a valve on the supply line.

The chamber T_A was not varied; it was maintained at a favorable value (30°C) in the light and near the temperature of the chilled water bath (25°C) in the dark. During a 75-minute simulated sunrise and again after lights were turned off, there was a transition between these two temperatures. After the temperature transition, T_A was held constant (within 0.25°C) at all heights within the chamber, except directly in front of the blower (1°C lower). For the greenhouse experiment, T_A ranged from 20 to 30°C.

Air flow. The recirculating cool air in the chamber absorbed the heat load from the lamps and also maintained a low R_A. Air velocity was sufficient to cause some visible leaf flutter but not buffeting. In addition to the recirculation of air within the chamber, airflow was metered through the chamber for measuring transpiration by the change in water vapor concentration as it passed through the chamber. Suction for the metered airflow came from a blower attached downstream from the chamber. A turbine meter was used to measure the flow rates, and a dew-point hygrometer was used to measure the vapor density difference by monitoring the reference air and the chamber air.

Measurement of R_A in the chamber and calculation of R_L. A low chamber R_A is required for stomatal closure and the consequent increase in R_L to correlate highly with reduced transpiration. R_A was measured with a simulated cotton plant whose leaves were made of green blotting paper embedded with 0.1-mm copper–constantan thermocouples. The following equation was used:

$$R_A = \frac{\Delta d_v}{E} \qquad (2)$$

where Δd_v is the difference in vapor density between the blotter leaf and the ambient air (g cm^{-3}) and **E** is the evaporation rate from the blotter (g cm^{-2} s^{-1}). In calculating R_A in this manner, the assumptions are that the evaporating surface is saturated with water vapor, and water loss from the blotter must overcome only an external resistance (in contrast to water molecules inside a living leaf, which must overcome both stomatal and boundary layer resistances). If these assumptions are valid, a measurement of the blotter temperature will give the saturation vapor pressure (from tabular values). Other necessary data are ambient vapor pressure and E. These were measured with a dew-point hygrometer by alternately measuring the dew point of the reference and chamber air. Then the difference in dew-point temperatures caused by the simulated plant in the chamber was converted to vapor density (g cm^{-3}) and multiplied by the bulk flow of air through the chamber (cm^3 s^{-1}) to give E.

The value of R_A is required as part of the data for calculating R_L from chamber measurements. To obtain R_L, equation (1) is rearranged:

$$R_L = \frac{\Delta d_v}{T} - R_A \qquad (3)$$

Since not only wind velocity but also leaf dimensions (both parallel and at right angles to the wind direction) influence R_A, the use of large leaves like those of cotton (15 × 14 cm) for measuring R_A in the chamber is a compromise when applied to plants like mesquite and jojoba with smaller leaves. Nevertheless, the value of R_A for the chamber (0.9 s cm^{-1}) is low enough so that only a slight error in calculated R_L values would occur even if the true R_A were as low as 0.1 s cm^{-1}.

Illumination. To simulate a sunrise, the bank of 24 400-W mercury vapor lamps was turned on sequentially by a card programmer. They were located 1.5 m above the chamber and provided an irradiance of 0.5 ly min^{-1} (an illuminance of 44 klx) at midlevel in the chamber. With appropriate positioning of the plant, most of the leaves were at heights above the midlevel point and thus were exposed to saturation illuminance for stomatal opening (Ehrler and Van Bavel, 1968). Thus, for well-watered plants, minimal R_L values could be expected.

Soil matric potential (ψ_M) Plants were grown in a standard soil mixture consisting of two parts each of peat moss and sand and one part of Avondale loam packed at a known bulk density in either 2-liter pots of acrylic plastic or 3.8-liter crocks. The soil was fertilized by periodic saturation with half-strength Hoagland nutrient solution, then desaturated at a known tension to a repeatable value of 0.39 volumetric water content (θ_v). A tensiometer in the soil at this θ_v read -4 cb. Such a tensiometer functioned adequately when only moderate soil water depletion was permitted. However, certain experiments were continued until transpiration had decreased 90 percent. This low rate did not occur until θ_v approached the wilting point (-15-bar percentage). Thus, during the latter part of the experiment, ψ_M values were lower than the -0.8-bar limit of a conventional tensiometer. Therefore, a calculated ψ_M was substituted for the measured value. ψ_M was obtained by calculating θ_v and referring it to the soil water characteristic curve, the assumption being that the soil water was depleted homogeneously. θ_v was obtained by multiplying the weight (θ_w) of the soil water content by the bulk density of the soil. In turn, θ_w was calculated as a residual by subtracting from the initial soil–plant system weight the weights of (1) the pot tare, (2) the oven-dry soil, and (3) the plant fresh weight (obtained immediately at the end of the experiment). After calculation of the initial ψ_M, half hourly transpiration values were used to indicate the subsequent soil water depletion and consequent lowering of ψ_M.

Leaf temperature (T_L). For measuring T_L, butt-welded copper-constantan thermocouples 0.1-mm in diameter were inserted in the leaves of jojoba and agave. In mesquite, the thermojunction tended to slip out of the mesophyll of a leaflet and therefore was placed in the rachis of a four-pinnate leaf. Test results with this type of thermojunction agreed to within 1°C with those obtained with a 0.05-mm thermojunction touching the lower epidermis of the leaflet immediately adjacent to the 0.1-mm junction in the rachis, in an environment similar to that of the present experiment.

Relative leaf water content (RLWC). Jojoba leaf hydration was measured continuously by monitoring the RLWC.

$$RLWC = \frac{100(W_f - W_d)}{W_s - W_d} \tag{4}$$

where W_f, W_s, and W_d are the weights of a known leaf area when fresh, i.e., at any time during the test, fully turgid (RLWC = 100), and oven-dried, respectively. A beta-ray gauge (Nakayama and Ehrler, 1964) was placed on a representative upper leaf. The greater-than-normal density–thickness (D–T, the weight per unit area) of jojoba leaves (40 to 70 mg cm^{-2}) made it necessary to use ^{204}Tl instead of ^{99}Tc. The signal from the gauge was fed to a scaler, rate meter, and strip-chart recorder to give a continuous record of leaf thickness (counts per minute as raw data). Later these data were calibrated in terms of equivalent D–T (mg cm^{-2}), which represented W_f in equation (4). At the end of the experiment the intact plant was kept in darkness in a tent at saturation vapor pressure to permit recovery from the previous dehydration. To obtain W_s, rather than use the destructive method of cutting out a disk and floating it on water (Jarvis and Slatyer, 1966), the rehydrated leaf was counted again with the beta-ray gauge after having been in the tent. Finally, W_d was measured after cutting out the leaf area circumscribed by the gauge and oven drying it.

Greenhouse environment. Midday solar radiation while corn and agave were growing outside the greenhouse averaged 1.0 ly min^{-1} (82 klx). To decrease the load on the evaporative coolers, a netted fabric shade screen was placed on the outside of the glass. This shade reduced radiation values inside the greenhouse about 50 percent. Thermostated steam radiators and evaporative coolers were used to hold the T_A range to from 20 to 30°C. The vapor pressure ranged from 12 to 17 mb and was prevented from dropping to low levels with an evaporative cooler regulated by a humidistat.

RESULTS AND DISCUSSION

Chilean Mesquite

Illuminance. As is true for most plant species, the stomates of Chilean mesquite close in the dark and open in the light (Fig. 1). When illuminance was varied (the saturation deficit held at a mean value of 27.4 mb, air temperature at 30°C, and Ψ_M at −4 cb), the equilibrium transpiration increased from 21 g m^{-2} h^{-1} in darkness to 185 g m^{-2} h^{-1} at 22 klx, and 230 g m^{-2} h^{-1} at 44 klx. Corresponding R_L values were 30.3, 2.8, and 1.8 s cm^{-1}, respectively. That the transpiration curve did not level off might be interpreted as a failure of mesquite to achieve saturation illuminance. However, the small decrease in R_L between 22 and 44 klx, combined with the low absolute value attained (1.8 s cm^{-1}), suggests that a minimal R_L was reached. Of eight different species tested in this and other experiments, the R_L value was among the lowest attained in the chamber, and was second only to sunflower (1.3 s cm^{-1}) under the same conditions. With its high transpiration rate, Chilean mesquite ranks among the high water-using species, provided soil water is easily available.

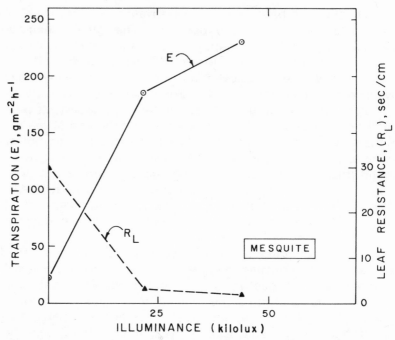

FIGURE 1   Effect of illuminance on the transpiration rate and leaf resistance of Chilean mesquite at a mean saturation deficit of 27.4 mb and a soil matric potential of −4 cb.

Saturation deficit. When illuminance was held constant at 44 klx and ψ_M at −4 cb, an increase in the saturation deficit (SD) caused a linear increase in the transpiration rate over a SD range from 15.5 to 27.4 mb (Fig. 2). This resulted in the high rate of 252 g m^{-2} h^{-1} at a SD of 27.4 mb. Although R_L was not quite as low as in the experiment when illuminance was varied, it was low enough to indicate open stomates. The linear regression constants for this relation were applied to a typical field environment for Phoenix, Arizona, in mid-June (T_A of 40°C; vapor pressure, 9 mb). This calculation yielded the phenomenal transpiration rate of 750 g m^{-2} h^{-1}. This potential rate is about three times the rate attained at a SD of 27.4 mb, which already ranks high when compared to values in the literature. Of course, this predicted rate is valid only for a SD of 64.8 mb. Actually, as the transpiration rate accelerated, the increased amount of evaporative cooling would tend to lower T_L, which, in turn, would decrease the transpiration rate below the initially predicted value. Nevertheless, this example illustrates the tremendous gradients for loss of water vapor in desert environments, and shows the urgency for plants to exert maximal resistance to the escape of water vapor from leaves when soils are dry.

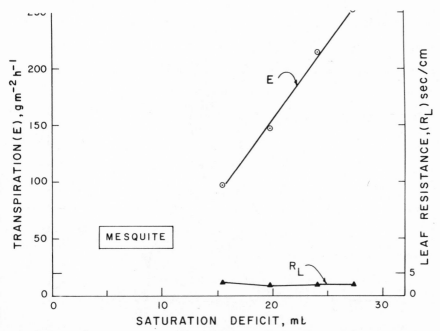

FIGURE 2 Effect of the saturation deficit on the transpiration rate and leaf resistance of Chilean mesquite at an illuminance of 44 klx and a soil matric potential of −4 cb.

Drought. Chilean mesquite plants were subjected to a steady soil water depletion by continued transpiration while the following environmental factors were held constant: 44 klx illuminance, 20 mb SD until all the lights were on and then about 26.9 mb (until the end of the experiment), and −6 cb ψ_M initially. In the dark, R_L was 32 to 36 s cm^{-1} (Fig. 3), consistent with previous results and with the low transpiration rate (about 17 g m^{-2} h^{-1}). After the simulated sunrise, transpiration increased to more than 200 g m^{-2} h^{-1}, but remained constant near this level only for 1.5 hours before decreasing gradually.

The decrease coincided with the point at which R_L began increasing from minimal values of about 2 s cm^{-1}. About the same time ψ_M began decreasing (becoming more negative) from the well-watered soil value of −6 to −1500 cb (the conventional soil wilting point) before increasing in response to irrigation. The effective transpiration control is shown by a gradual increase in R_L under strong illumination to a high of 47 s cm^{-1}, higher than a typical value for darkness, and the simultaneous decrease in transpiration to 18 g m^{-2} h^{-1} (a 91 percent decrease from the peak rate). A 5-minute irrigation restored the total transpiration loss from the

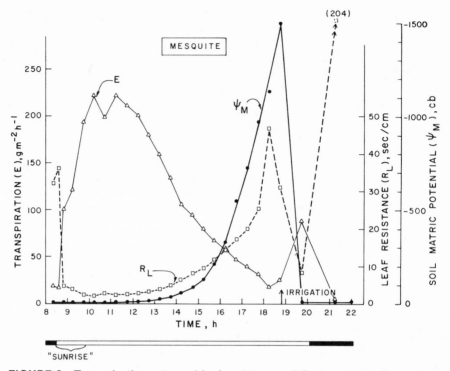

FIGURE 3 Transpiration rate and leaf resistance of Chilean mesquite and soil matric potential as a function of time after irrigation at illuminance levels from 0 to 44 klx and a mean saturation deficit of 26.9 mb.

beginning of the experiment and caused the expected increases in ψ_M and transpiration and decrease in R_L.

When the lights were turned off, transpiration decreased to a minimal value as R_L increased to 204 s cm^{-1}, which was probably a transient phenomenon. These experiments demonstrate the utility of the transpiration equation in accounting for the interrelations of environmental factors upon transpiration, and simultaneously emphasize the extreme importance of R_L in regulating water loss.

Jojoba

Drought. At the start of the experiment the soil was well watered, with ψ_M at -6 cb (Fig. 4). The 100 percent RLWC demonstrated that the plant was fully hydrated. Rather surprisingly, R_L values in the dark were not high, ranging from 10 to 17 s cm^{-1} (Figs. 4 and 5) as compared with 30 to 36 s cm^{-1} for mesquite (Figs. 1 and 3). As illumination increased, transpiration rapidly increased to 64.5 g m^{-2} h^{-1} at 44 klx. The rate remained

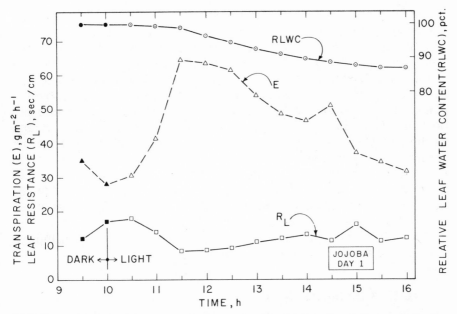

FIGURE 4 Drought effects on the transpiration rate, leaf resistance, and relative water content of jojoba at illuminance levels from 0 to 44 klx and a mean saturation deficit of 21.6 mb: day 1.

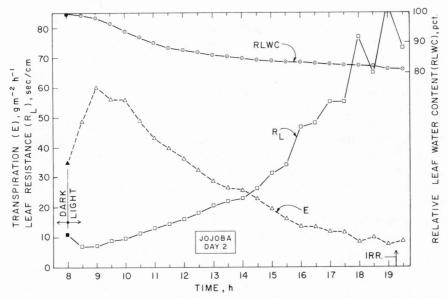

FIGURE 5 Drought effects on the transpiration rate, leaf resistance, and relative leaf water content of jojoba at illuminance levels from 0 to 44 klx at a mean saturation deficit of 22.9 mb: day 2. The first data point in each curve represents a value in darkness.

above 60 g m^{-2} h^{-1} for only 1.5 hours, after which it gradually began decreasing, except for a spurt near 1430. This decrease was correlated with the point at which the RLWC decreased below 95 percent. Simultaneously, R_L just started to increase from its minimal value of 8 s cm^{-1} in the light. By the next day (Fig. 5), the RLWC had returned to 100 percent, apparently from an overnight redistribution of water within the plant. Transpiration accelerated with more illumination, reaching a peak at 0900 of 60 g m^{-2} h^{-1} (almost the same level as on day 1). Again, as the RLWC decreased below 95 percent, transpiration began to decrease and R_L to increase. This decrease in RLWC and transpiration and increase in R_L progressed steadily until 1915, when the RLWC had decreased to 81 percent, the transpiration rate to only 8 g m^{-2} h^{-1} (an 87 percent reduction from the peak rate), and R_L had increased to about 90 s cm^{-1} (far above the value attained in darkness).

These data demonstrate jojoba's great sensitivity to decreasing hydration and the effectiveness of R_L in decreasing transpiration.

Agave Contrasted to Corn

In all the experiments with agave and corn, drought effects were avoided by an irrigation schedule where ψ_M ranged from -3 to -20 cb. During 12 hours of darkness the transpiration of corn was only about 7 percent of that for an equal period of light, whereas for agave it was 200 percent (Table 1). Thus, agave shows an effective resistance to the escape of water vapor in the light and a lowered resistance in the dark; both responses show an inverted rhythm of stomatal response as compared with nonsucculent plants like mesquite, jojoba, and corn.

On a diurnal basis (Fig. 6), the transpiration of corn in a greenhouse paralleled the evaporative demand and illuminance level (as shown by the response of corn transpiration and water loss from Piche† evaporimeter disks during midday, when light was adequate). In contrast, agave in the same greenhouse environment had an hourly transpiration rate almost independent of the evaporative demand at a low rate as compared with the high peak rates attained by corn.

This large difference in transpiration rates was reflected in T_L values (Fig. 7). Near noon, while T_A was 30°C, T_L for agave was more than 8°C higher than that for corn, which was 1.5 to 2°C higher than T_A. In the dark (Fig. 8), T_L for both plant species was near T_A; however, T_L tended to be slightly but definitely lower in agave as compared with corn, although the agave leaf was several times thicker than the corn leaf.

---

† Trade names and company names are included for the benefit of the reader and do not indicate endorsement or preferential treatment of the product listed by the U.S. Department of Agriculture.

TABLE 1 Transpiration rate of corn and agave in light and darkness for 2 days

Pot	In 12 h of light Day 1	Day 2	In 12 h of darkness Day 1	Day 2
	Corn transpiration ($g\ m^{-2}\ h^{-1}$)			
1	80.7	71.5	5.40	3.26
2	101.2	87.8	7.83	5.92
3	106.6	92.5	7.44	5.90
$\overline{X}_{3\ pots}$	96.2	83.9	6.89	5.03
$\overline{X}_{2\ days}$	90.0		5.96	
	Agave transpiration ($g\ m^{-2}\ h^{-1}$)			
1	22.5	11.4	43.8	22.5
2	4.69	7.31	36.4	19.9
3	32.0	9.70	29.2	28.0
$\overline{X}_{3\ pots}$	19.7	9.47	36.5	23.5
$\overline{X}_{2\ days}$	14.6		30.0	

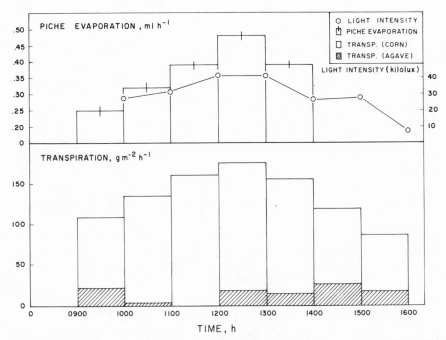

FIGURE 6 Hourly transpiration rate of corn and agave as affected by the evaporative demand in the greenhouse.

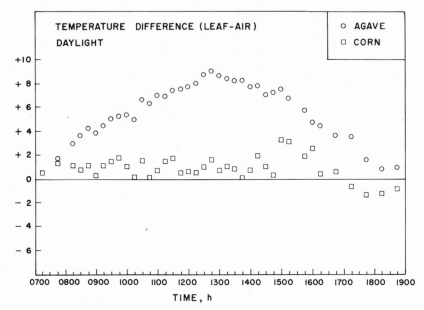

FIGURE 7   Daylight values of the temperature difference between leaf and air in corn and agave.

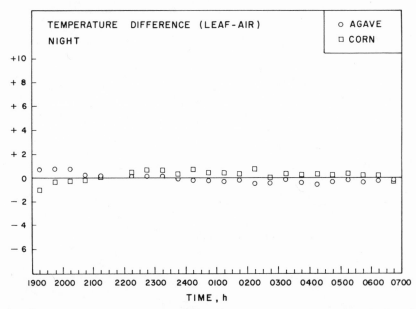

FIGURE 8   Nocturnal values of the temperature difference between leaf and air in corn and agave.

Values of R_L measured with a leaf resistance meter were very high for agave in the light and moderately low in the dark (Table 2). The long transit times required to make the measurements in the light cast some doubt as to the accuracy of the absolute value, i.e., about 200 s cm^{-1}. However, R_L undoubtedly was very high in the light, indicating tightly closed stomates, and lower in the dark, indicating a lower resistance to the outward diffusion of water. Moreover, the upper epidermis of agave, with a higher stomatal frequency (4850 stomates per cm^{-2} as compared with 3717), had a much lower R_L than the lower epidermis, which indicates the good sensitivity of the resistance meter.

In corn, R_L decreased to a minimum of 7.1 s cm^{-1} in the light as compared to 93 s cm^{-1} in the dark; both values were consistent with the relative transpiration rates. Also, as in agave, the epidermis with the higher number of stomates had the lower R_L (in the light). If the illumination had been at least 65 instead of 35 klx, the R_L of corn would have been much lower, probably 2 s cm^{-1} (Ehrler and Van Bavel, 1968). Thus, the transpiration rates, leaf–air temperature differences, and R_L values were consistent in indicating stomatal opening in the light and closure in the dark for corn, with the opposite cycle for agave, a succulent. This is strong presumptive evidence for the functioning of crassulacean acid metabolism in agave.

CONCLUSIONS

Transpiration was effectively restricted for agave even when the plant was well watered; the R_L values did not decrease below 17 s cm^{-1} during its period of maximal stomatal opening, and increased to above 200 s cm^{-1} in the light. Mesquite and jojoba also drastically restricted transpi-

TABLE 2 Measured leaf diffusion resistance in corn and agave as related to stomatal frequency in the upper and lower epidermal layers

	Epidermis		
	Upper	Lower	$\bar{x}$
Leaf diffusion resistance (s/cm)			
Agave			
Light	194.0	292.0	239.0
Darkness	15.2	19.6	17.1
Corn			
Light	8.5	6.1	7.1
Darkness	88.0	100.0	93.0
Stomatal frequency (stomates/cm²)			
Agave	4850	3717	
Corn	5077	9247	

ration when the soil was dry, developing R_L values of 47 and 88 s cm^{-1}, respectively, when the transpiration rate had decreased 90 percent from the potential rate. Despite this similarity between mesquite and jojoba when the soil was dry, there was a pronounced difference between species when the soil was wet: under high illuminance R_L was 8.3 s cm^{-1} for jojoba, but only 1.8 s cm^{-1} for mesquite. These differences cannot be attributed to a lack of stomata on one epidermis, because measurements of R_L with a leaf resistance meter (Van Bavel et al., 1965) showed that both mesquite and jojoba are amphistomatal. Perhaps jojoba utilizes one or more of the same mechanisms for developing high R_L values as agave, such as low stomatal frequency, sunken stomata, and an extremely impermeable cuticle. However, that such mechanisms are used must be confirmed by microscopic measurements.

LITERATURE CITED

Ehrler, W. L., and C. H. M. Van Bavel. 1968. Leaf diffusion resistance, illuminance, and transpiration. Plant Physiol. 43:208–214.

Gates, D. M. 1968. Transpiration and leaf temperature. Ann. Rev. Plant Physiol. 19:211–238.

Jarvis, P. G., and R. O. Slatyer. 1966. Calibration of β gauges for determining leaf water status. Science 153:78–79.

Nakayama, F. S., and W. L. Ehrler. 1964. Beta ray gauging technique for measuring leaf water content changes and moisture status of plants. Plant Physiol. 39:95–98.

Van Bavel, C. H. M., F. S. Nakayama, and W. L. Ehrler. 1965. Measuring transpiration resistance of leaves. Plant Physiol. 40:535–540.

COLORATION AND ITS THERMAL CONSEQUENCES FOR DIURNAL DESERT INSECTS

William J. Hamilton III

*Division of Environmental Studies, University of
California, Davis, California*

Abstract

Behavioral thermoregulation by certain desert insects permits elevated body temperature levels to be maintained for a maximum interval. Since the metabolic rate of poikilotherms rises with body temperature, these tactics may permit maximum growth and reproductive rates. In the case of the diurnal desert tenebrionid beetles considered here, the level of elevated temperatures is about the same for most species. The lethal temperature is about 48°C. At body temperatures below 38°C, these beetles behave so as to increase body temperature. Between 38 and 40°C, behavior is modified to maintain body temperatures at that level. Above 40°C, behavioral thermoregulation and morphology prevent a further rise in body temperature. Maxitherms attempt to maintain this elevated thermal level as long as possible, and do so for considerable intervals.

When coloration is primarily an adaptation to heat exchange, the beetle species that are the focus of this study adopt either black or white coloration. It is suggested that black and white coloration is an adaptation to maximize the mean interval of potential activity at elevated body temperatures.

INTRODUCTION

In this paper I shall discuss the thermal consequences of coloration to a desert animal. However, any consideration of the adaptive significance of coloration must also include other adaptive values. Elsewhere (Hamilton, 1973) I have reviewed the significance of all coloration phenomena to life, and have concluded that most, if not all, animal coloration can be assigned an adaptive value as communication, crypsis, or optimizing radiation relationships. The surfaces of most larger desert animals are notable for their dull coloration, a phenomenon that agrees with the muted tones of the desert environment and emphasizes the role of coloration in securing camouflage. The physical and biological principles underlying camouflage, especially for lizards, have been reviewed by Norris (1967). Visual communication (within species) by desert animals may be enhanced by colored surfaces. However, most desert animals confine colored communication surfaces to the underside of the body or to skin folds; thus the conflicting requirements of camouflage and communication are both served.

The hypothesis which is the framework around which I will evaluate the thermal significance of coloration for desert animals is that certain animals are maxitherms (Hamilton, 1973); i.e., they maintain body temperatures as high as possible for as long as possible. It is argued that the maximum level is determined by the limits of potential evolutionary adaptation, which are approximately the same for all multicellular animals. For these animals, upper lethal limits are in the 47 to 52°C range, and the preferred body temperature is 37 to 41°C. The hypothesis predicts the following:

1. The upper lethal body temperature of maxitherms will, except for physiological adaptation, be essentially the same regardless of the environment in which they live. This prediction is based upon the proposed hypothetical limits of evolutionary adaptation.

2. Differences in the lethal limits of various maxithermic species will reflect the recent thermal experience of respective populations, and especially the degree to which they have been able to maintain maximum body temperatures.

3. When ambient thermal conditions rise in the middle of the day above tolerable levels, activity of diurnal species will cease and there will be a retreat to more moderate thermal environments. In such an environment, optimum conditions will occur twice daily, once as temperatures are rising in the morning and again as they are falling in the afternoon or early evening.

4. When activity cycles are bimodal on some days and unimodal on others, the bimodal days will be warmer, and the first burst of activity on the bimodal days will come earlier than the single crest on unimodal days.

5. When a species occurs in several environments with different thermal characteristics, the periodicity of activity will differ from place to

place, which enables each local population to establish optimum thermal relationships to local conditions.

6. Where more moderate conditions prevail and ambient conditions are such as to permit maintenance of body temperatures below 41°C, activity will continue through the middle hours of the day.

7. When sufficient energy is available, body temperatures of individuals and the population will rise to the 37 to 41°C range. Body temperatures will be maintained in this range by behavioral thermoregulation relative to the morphology of each species until it is no longer possible to resist further heat gain.

Gates (1974) has argued that analysis of the thermal relationship of animals to their environment is a highly quantitative science and he implies that only quantitative evaluations are of value. However, such an approach has limited biological application, especially if it presents only quantitative measures of heat flow. We can ultimately develop quantitative models (e.g., Porter and Gates, 1969) that reveal how an animal responds to its meteorological environment and, eventually, how an animal responds physiologically and behaviorally to that environment. These models are useful, and when they include detailed measurements of the physical parameters of the environment in which animals live, they may help us to determine the nature of animal adaptation to the environment. But mobile animals are able to choose their environment, and most terrestrial animals have available to them a broad spectrum of physical environments and thermal conditions. The evidence developed relevant to the hypothesis evaluated here is both quantitative and qualitative. Throughout this review my emphasis will be upon the actions of free-ranging animals and the interpretations of their actions relative to their coloration and ambient microclimatic conditions. In this analysis most critical interpretations are based upon comparative qualitative data.

STUDY AREA AND METHODS

The investigations reported here were made principally in the Namib Desert, from 20 km south of Gobabeb, South West Africa, to Mocamedes, Angola, and eastward to more mesic environments throughout this desert. Three distinct physiognomic substrates in the Namib were investigated: (1) sand dunes, (2) gravel plains, and (3) dry river bottoms. Physically and biologically, the sand dunes are the outstanding feature of the Namib. These largely vegetationless sand masses extend from Luderitz in South West Africa to the Coroca River in Angola, and eastward up to 80 km from the Atlantic coast.

Activity Cycle Censuses

The Namib Desert dune system is occupied by a great diversity of endemic species and genera of tenebrionid beetles (Koch, 1961, 1962). In sparsely vegetated environments such as desert dunes, observations of the periodicity and nature of activity of these insects is facilitated by the lack of vegetation. Details of census techniques are provided elsewhere (Hamilton, 1971). The procedure involved repeatedly walking traverses along predetermined routes and noting the number of active insects and their behavior.

Many of the activity cycle data presented here are based upon the record of a single observation date, because these beetles are highly responsive to the varying ambient conditions that influence microclimatic temperature conditions. Since these conditions change from day to day and from hour to hour, pooled data for several days would increase the apparent temporal variability of activity responses. Furthermore, a route can be established that will intercept reasonably large numbers of insects. Since the actions of these individuals are independent from one another, the activities of individuals encountered on each census traverse samples the actions of the population from which they are drawn. Furthermore, it is assumed that, insofar as there are significant differences among the activity cycles of the same species in different habitats, the differences can be assigned to microclimatic differences among the habitats. A less confident assumption is that differences in activity cycles between species can be assigned to differences in morphology and ambient microclimate. However, this assumption lies at the heart of the general hypothesis proposed here and elsewhere (Hamilton, 1973), and for purposes of this discussion it is accepted as true.

All time measurements reported are local sun times. These values permit direct comparison of the activity responses of populations widely separated in space within a single time zone, as in South Africa and South West Africa, where local standard time spans nearly 3 hours of sun time.

Body Temperature Measurements

Accurate measurements of the body temperature of free-ranging desert beetles have not previously been reported. By comparison, there are now tens of thousands of measurements of reptile body temperatures, most of them secured by capturing a specimen and inserting a thermistor or mercury thermometer probe into the cloaca. The tight armature of desert beetles does not make them susceptible to such penetration. The procedure adopted here was to capture the insect, hold it by the legs, remove the head by pinching it off, and insert a thermistor probe into the center of the remaining body cavity.

Weather Measurements

To reduce the limitations of conventional weather instrumentation, a system for determining localized microclimatic conditions was devised. Spot measurements were made after each beetle body temperature was determined. The process was facilitated by the use of rapid response thermistors, which come to equilibrium in a matter of seconds. These little probes can be inserted into small places, and in soft sand they can be pushed into the substrate to monitor subsurface conditions.

For all surface active diurnal insects, the most significant thermal feature of the environment is ambient temperature. In most observations body temperatures exceed ambient air temperatures, usually by several degrees. Ambient air temperatures establish a minimum body temperature for small desert animals. This is particularly important to keep in mind when the significance of radiative heating is considered relative to thermal equilibria. Air temperatures were determined with a small, shaded thermistor placed in the exact place where the beetle was captured.

Wind speed was measured with a hot wire anemometer. Six readings were taken at 10-second intervals after measurement of body temperature, and their values averaged.

RESULTS AND DISCUSSION

Solar Radiation and Coloration

Incoming solar radiation is differentially reflected by animal surfaces. We are able to see about half of incoming solar energy as visible light, i.e., as color. In my opinion, the visible coloration (reflectance) of pelage or surface accurately defines the relative position of invisible reflectance in the ultraviolet and near infrared for most animal surfaces. Gates (1974) has argued that this is not the case, and he may be correct for many inanimate surfaces. But for desert birds (Fig. 1) and insects (Fig. 2) the visible accurately predicts the relative total reflectance of animal surfaces. Thus, the white pelican is whiter in most of the infrared and ultraviolet than are black and camouflaged (sparrow) birds, and the pelican's total radiant energy relationship to the environment is more reflective to radiant energy than the black and camouflaged birds with which it is compared. Norris's (1967) reflectance curves for the lizards *Callisaurus draconoides* and *Holbrookia maculata* indicate that the

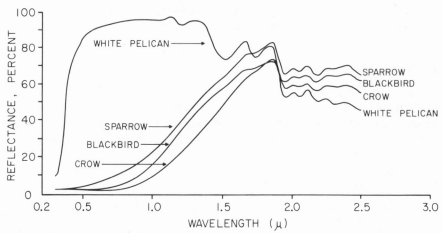

FIGURE 1 Dorsal surface feather reflectance of a white bird, the white pelican, *Pelecanus erythrorhynchos,* a camouflaged brown sparrow, *Zonotrichia leucophrys,* and two black birds, the Brewer's blackbird, *Euphagus cyanocephalus,* and the crow, *Corvus brachyrhynchos.*

visibly paler ventral surface is also paler throughout the measured part of the infrared spectrum. For the tenebrionid beetles considered here, the visible also predicts species differences in the relative position of the infrared and ultraviolet (Fig. 2). Thus, in most cases we can conclude that visible coloration accurately predicts the general nature of adaptation, whether it emphasizes optimization of thermoregulation, communication, or camouflage.

Since color is a function of reflectance, it follows that animals of equal size but different color will develop different body temperatures under similar ambient conditions. Such differences may be offset if high wind velocities enhance rates of convective cooling so that there is a negligible temperature excess, if a large part of the unreflected energy is transmitted through the body, or if physiological mechanisms of cooling were more highly developed in one color form than in the other. These possibilities are considered next.

As wind speeds increase, the temperature excess of beetles decreases. This variable can be identified, in the case of free-ranging beetles, with the crude methods used here to monitor body temperature and wind. For example, despite great variation, the mean body temperature of the large black beetle, *Eustolopus octoseriatus*, ranges from an average temperature excess of 3.6°C to less than 1°C at wind speeds in excess of 4 m/s (Table I). This is true for black and for white beetles. At high wind velocities, no significant temperature excess can be measured. But if these beetles are acting essentially as inanimate physical bodies, as is suggested here, a slight unmeasured temperature excess

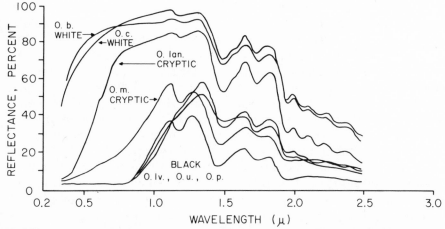

FIGURE 2 Dorsal abdominal surface reflectance of two white Namib Desert beetles, *Onymacris bicolor* (O. b.) and *O. candidipennis* (O. c.), two camouflaged beetles, the golden *O. langi* (O. lan.) and the brown striped *O. marginipennis* (O. m.), and three black beetles, *O. laeviceps* (O. lv.), *O. unguicularis* (O. u.), and *O. plana* (O. p.).

should remain at any wind speed. High average wind velocities will reduce the absolute value of white or black coloration, but white surfaces are more reflective than black ones under all ambient conditions.

Physiological evaporative cooling and heating by metabolism are of minor importance for small beetles (Edney, 1971).

There is essentially no transmission of solar radiation through black cuticles, and any energy passing through the white dorsal surface of the white Namib tenebrionids will be intercepted and converted to heat by the brown lining on the underside of the elytra or by the black underside of these animals before passing completely through the body. Norris (1967) also concludes for diurnal desert reptiles that transmission is

TABLE 1 Body temperature excess above ambient for free-ranging individuals of the black tenebrionid beetle, *Eustolopus octoseriatus* at ranked wind speed intervals

Wind speed (m/s)	Temperature excess (°C)	Range temperature excess	n
0–1.2	3.6	0.1–5.4	16
1.3–1.8	2.7	0.1–7.9	22
1.9–2.4	2.2	0.6–6.0	16
2.5–3.0	1.7	1.3–5.1	7
3.1–3.6	1.4	0.2–2.1	12
3.7–4.2	0.8	0.2–1.3	11

negligible. In any case, transmitted energy can only magnify the difference between the black and white species. Thus, it is reasonable to assign any difference in temperature excess between black and white beetles of the same size and general shape to differences in reflectance.

Previous workers have not always assumed that black coloration actually exaggerates rates of heat gain. Buxton (1924) seemed to have established this by a simple experiment. He tethered two differently colored morphs of the desert grasshopper, *Calyptamus coelesyriensis*, in sunlight. The dark, nearly black morph reached a temperature 4.5°C warmer than that of the "desert-colored" pale morph. However, Pepper and Hastings (1952) did not find similar differences in morphs of another grasshopper, *Melanoplus*, and other recent reviews of the heat exchange of desert animals have minimized (Digby, 1955; Edney, 1967) the relevance of visible coloration to animal heat exchange. Comparisons of the heat-exchange rates of dimorphic insects such as grasshoppers (Buxton, 1924; Pepper and Hastings, 1952) may be deceptive because the tacit assumption that the morphs are alike but for color is not correct. The dark morph of the migratory locust, *Schistocerca gregaria*, for example, differs from the light phase with respect to external (Dudley, 1964) and internal (Ellis and Carlisle, 1961; Nolte, 1964) morphology.

When body temperature differences are determined for different species in the case of the white *Onymacris brincki* and the black *O. rugatipennis* (Edney, 1971) and for black *Eleodes armata* beetles painted white (Hadley, 1970), the black beetles are significantly warmer than white ones. Since the reality of these differences has often been disputed, results of some additional field experiments are included here. Abdominal temperature measurements were made of live, tethered pairs of white *O. brincki* and black *O. unguicularis* from an area of sympatry. When exposed to identical ambient conditions in sunlight in an open courtyard, the black species developed a greater equilibrium temperature excess in all 20 experiments ($p < 0.001$, paired t test). Furthermore, these differences can certainly be assigned to surface albedo, as the differences vanish when the white elytra of white *O. brincki* are painted with a flat optical black paint (p = n.s.), or when the test subjects are shaded (p = n.s.).

The relationship of black coloration to heat exchange thus seems clear enough. Black coloration increases the rate of heat gain of desert poikilotherms when they are exposed to sunlight.

The interpretation to be assigned to these observations is not so obvious. Most commentators have made the tacit assumption that heat is a positive hazard to any desert animal and that any trait which promotes heat gain is disadvantageous to such animals. Perhaps this assumption is at the heart of earlier failures to solve the "black beetle puzzle," that desert animals in general, and desert beetles in particular, tend to be black. If heat gain conveys a positive advantage to certain desert animals, it may account for the adaptive significance and evolution of black desert coloration. The principal result of changes that accompany an increased body temperature are increases in metabolic and locomotion rates. The evolutionary advantages of sustaining a high body temperature are probably greatest relative to metabolic rate. As long as food supplies are adequate to sustain metabolism, a higher metabolic rate will result in a higher rate of conversion of food to energy, more rapid growth, and a higher rate of reproduction. Selection will favor these attributes, and adaptations to maximizing body temperatures are widespread among poikilothermic animals, including certain desert animals. Lizards, for example, are known to sunbathe, seeking heat until body temperatures have climbed to near lethal levels (Norris, 1967). Many species of lizards undergo a daily cycle in color change, becoming dark at dawn and blanching in the middle of the day when heat gain is no longer favorable. A similar adaptation has been noted by Key and Day (1954a, b) in the Australian grasshopper *Koscuiscola tristis*, emphasizing the reality of the positive adaptation of insect coloration to heat gain. The advantage of heat is time dependent, and since the rate of most biological processes more than doubles with each 10°C increase in temperature, it is possible to accrue a considerable advantage relative to other members of the same population by sustaining a higher body temperature for a relatively short period of time.

Deserts are concentrated at middle latitudes where sharp seasonal weather fluctuations occur, providing considerable potential advantage for heat-gain adaptations. Deserts are also characterized by sharp daily changes in temperature. Cold nights are the rule, another circumstance favoring heat-gain and heat-retention adaptations. In addition, deserts are characterized by a high degree of insolation, and thus provide a predictable and persistent source of radiation.

The preceding data and logic indicate that a case can be made for the role of black desert beetle coloration as a positive adaptation to maximizing heat absorption.

If black coloration is to provide an advantage in heat absorption, there must be adequate mechanisms for heat dissipation. Because of their small size and high surface to volume ratio, evaporation of water is an impractical cooling mechanism for beetles (Edney, 1967, 1971). Desert tenebrionid, curculionid, carabid, and cincindelid beetles tend to have fused sclerites, a morphological adaptation minimizing water loss (Hellmich, 1933). Behavioral and morphological adaptations for heat regulation thus become supremely important. The small size of most insects is an effective preadaptation to desert environments, because it permits the use of small bits of shade or vertical temperature gradients that would be useless to larger animals. For example, the small Namib Desert tenebrionid, *Stenocara phalangium,* avoids excessively hot substrate sands by climbing onto small quartz pebbles, some no more than 1 cm in diameter. These thermal refuges are defended against intruding conspecifics (Henwood and Hamilton, in preparation).

Burrowing is probably the most important behavior for heat regulation because it permits access to a broad range (Fig. 3) of ambient temperatures where conductive heat exchange may rapidly bring body temperatures to sand temperature. Another important behavioral adaptation for heat regulation is fractionation of the surface activity period into intervals, so that feeding, mating, and forays abroad for other purposes are made in sorties (Holm and Edney, 1973), the duration of which is inversely related to ambient temperatures and radiation intensity. These adaptations are used by black Sahara Desert beetles of the genera *Adesmia* and *Pimelia,* which dig shallow burrows into the slopes of dunes, digging deeper as the surface sands warm (personal observations). Such a burrow can be dug in a matter of minutes. Usually they are placed at the base of a plant, which holds the soil together and prevents collapse of the tunnel. But burrows may be made in any depression, even in the open sand where natural irregularities, camel

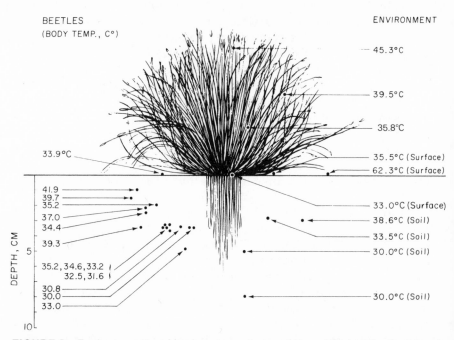

FIGURE 3   Environment and body temperatures of the white beetle, *Onymacris langi,* at midday. The deeper beetles may have been inactive on the date of these measurements.

tracks, or human footprints may provide the sloped surface for the start of a burrow.

Movement to shade is another behavioral adaptation for cooling that is facilitated by relatively small size. Bases of clumps of vegetation and, at the ends of the day, irregularities in the terrain may provide sufficient shade to permit cooling, which takes place very rapidly under such circumstances. Thus, behavioral temperature regulation centering about exposure and retreat from solar radiation is an important feature of the thermal ecology of black desert beetles.

The temperature of the desert, of both the surface sands and ambient air, frequently exceeds levels that, if attained as body temperatures, would be lethal. Yet heat death is probably not a hazard to desert beetles because they are able to escape from these environments when necessary. The energy required for thermoregulation is the energy required to move to and from places at different temperatures.

We know that white coloration is an available adaptation of desert beetles, because white species actually occur on the Namib Desert of South West Africa. The possibility that their white coloration may serve to reflect solar radiation has been rejected by the master investigator of this unique group of insects, C. Koch (1961). However, his reference to Bolwig's (1957) finding that the white tenebrionid species cannot withstand higher temperatures than the black species is irrelevant, as the upper lethal temperatures of most species that live in hot environments are nearly identical (Table 2). In any event, the suggestion that coloration differences secure differences in temperature tolerance levels is not implied. Rather it is suggested that white coloration may extend the activity time available to the white species in exposed environments.

Maximum Body Temperature Tolerances

Namib Desert tenebrionid beetles offer an exceptional opportunity to test the validity of the maxithermy hypothesis, because most of the diverse endemic species are restricted to specific macro- and micro-

TABLE 2   Maximum voluntarily tolerated body temperature and lethal temperatures (°C) for some diurnal black and white Namib Desert tenebrionid beetles (figures in parentheses are from Edney, 1971)

White species			Black species		
	Maximum voluntary	Lethal		Maximum voluntary	Lethal
O. bicolor	43.0	48.7	O. laeviceps	42.5	47.5 (48.5)
O. candidipennis	43.6	49.0	O. unguicularis	42.6	47.6
O. brincki	43.8	47.8	O. plana	42.7	(50)
O. langi	42.9	48.0	O. rugatipennis	43.8	(50)

climatic zones. Namib Desert climates range from cool Mediterranean conditions along the coastal bench to the hot extremes of the interior vegetationless dunes and gravel plains. The riparian forests along the perennially dry sand wastes of the river systems that transect the Namib provide intermediate conditions. Thus, if there were no upper limit to thermal evolutionary adaptation, the most heat tolerant species at each locality should be able to withstand and prefer higher or at least different temperatures than related species living at less extreme conditions. If, on the other hand, the most heat tolerant species in each environment are able to develop body temperatures that reach the limits of evolutionary adaptation to upper thermal extremes, then the lethal temperatures of various species living in quite different environments should be approximately the same, regardless of the environment. The results of interspecific comparisons (Table 2) indicate that, regardless of the environment, lethal and maximum voluntarily tolerated body temperatures are not significantly different.

Certain limitations to the data of Table 2 and the degree of confidence that can be placed in them need to be considered. The lack of statistically significant differences among populations, in the case of these species, may simply be related to sample sizes. The smaller the sample, the less likely it is that any real differences between populations will be detected. An adequate evaluation of similarities or differences among species will also need to consider physiological thermal adaptation. Even if the potential upper lethal tolerance is the same for a variety of species, significant differences may be detected due to differences in the recent thermal experience of each population. Thus, even though several species may attempt to maintain the same body temperature, one species may be able to do so more often or for a longer interval, raising its physiological adaptation to thermal stress higher than that of a species living in a region where opportunities for the development of maximum body temperatures occur less often or for shorter intervals.

The data for the maximum voluntarily tolerated body temperature can be interpreted only in a comparative, as opposed to a statistical, sense. The data are for the maximum single body temperature observation under the stated conditions. For these eight species, the value ranges only from 42.7 to 43.8°C. The values for *O. bicolor* and *O. candidipennis* are considerably higher than those previously reported (Hamilton, 1973), and are based upon extensive additional field observations in May 1974. The remarkable sameness of the maximum recorded voluntarily tolerated temperature for eight different species is, like the lethal temperature data for these same species, in agreement with the maxithermy hypothesis stated previously.

Intensive efforts were made to measure the body temperatures of these beetles during the hottest part of the hottest days. These data were accumulated under diverse climatic conditions during a period of nearly 1 year. They are thus not strictly comparable. In addition, the results include nearly 2000 measurements unevenly distributed among the eight component species (Table 2). The probability of measuring an extremely hot beetle of each species increases as sample size for that species increases.

Activity Rhythms

The maxithermy hypothesis assumes that every individual, and the population of which it is a part, is attempting to maintain approximately the same body temperature, somewhere between 37.0 and 41.0°C, for as long as possible. From this it follows that to be optimally adapted the activity cycle of every population of a species should vary according to (1) the environment in which it lives and (2) the ambient thermal conditions on any particular day. A species living mainly in the river bottom, *Onymacris rugatipennis,* provides a test of this hypothesis. Local populations live in a spectrum of local habitats. These habitats near Gobabeb, South West Africa, are distinctive in terms of exposure to radiation and the extent of shade cover. *Onymacris rugatipennis* is a large, highly mobile species, and individuals may range widely, traveling several meters per minute. Thus, they can, and occasionally do, range from more to less exposed habitats. But the habitats selected for special study were relatively extensive, and marked individual beetles were observed to remain within each of them for a period of days. There was relatively little exchange of individuals between these rather different environments during the brief period of the observations reported here.

The most protected thermally of these environments is the *Acacia* forest on the banks of the perennially dry Kuiseb River bottom. In most places occupied by *O. rugatipennis,* the canopy of this forest is broken, and patches of sunlight penetrate regularly to the leaf- and rubble-littered sandy substrate. An intermediate environment in terms of thermal amplitude is provided by the thorny bunch grass, *Stipagrostis,* and scattered annuals and perennials, especially *Nicotiana,* in the sand wash at midstream. There is no overhead woody vegetation in these environments, and the only cover available above the surface is provided by these forbs. The sand beneath the *Stipagrostis* clumps is used as an overnight retreat by the resident *O. rugatipennis* population. When the beetles living there are above the surface and active, they may rest in the complete shade under these clumps, in the filtered shade near the periphery of the vegetation, or range onto the open sands between the isolated patches of vegetation. The most exposed environment is on the sand piles collected about clumps of the naras bush, *Acanthasicyos horridus,* a thorny, green-stemmed, leafless, creeping perennial vine. Unlike the other two environments, the substrate sands here are sloped; and in the morning the sunward slopes warm more rapidly than any part of the other environments. *Acanthasicyos* occurs in isolated, widely spaced clumps that extend well out onto the gravel plains south of the Kuiseb River bottom. Individual *O. rugatipennis* inhabiting these clumps range out onto the completely unprotected sand-covered gravel plains to forage. The combination of the exposed sloped sands and the unprotected surrounding space made this the least protected environment inhabited by *O. rugatipennis* at Gobabeb.

In other parts of the Namib and even elsewhere at Gobabeb the degree of exposure to desert climate may be greater than it is in the *Acanthasicyos* thickets. But the environments described here had the commanding advantage of being susceptible to census on the same day and in rapid succession. The results of such a census for two dates, October 26, 1967, and January 2, 1968, are indicated in Figure 4. In the least exposed environment, the *Acacia* woodland, the peak activity level for the population may not be reached until well after noon. On warmer days, the activity in this environment becomes bimodal, with a crest in the morning and one again in the afternoon. In the moderately exposed environments, activity is bimodal, with the maxima shifting apart and toward the end of the day. In the most exposed environment, activity is shifted well toward the ends of the day. These differences among environments occur within 300 m of one another. Standard meteorological equipment placed in the sun in each of these environments revealed essentially no difference from site to site. Where thickets of the forb

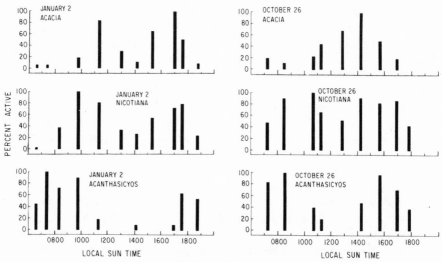

FIGURE 4 Surface activity of black beetles, *Onymacris rugatipennis,* in three different habitats on days of differing thermal characteristics.

Nicotiana glauca are well developed, individuals remain active throughout the middle hours of the day; and on colder winter days a single activity peak occurs. Under the dense foliage of the evergreen acacias, *Acacia albida* and *A. giraffae,* the peaks in activity converge on hot days; and during cool weather a single activity peak occurs.

These variations within each species complicate any attempt to relate activity cycles to coloration. If white coloration is an adaptation enabling a species to optimize thermal relationships to the environment, midday activity should be enhanced, as it is clear that maintenance of an early morning 38°C body temperature in late afternoon will be impossible as a result of the high reflectance and declining incoming solar radiation. To evaluate this point, I selected specific local sites where white and black members of the diurnal adesmine tenebrionids occur in the same habitat.

The first comparison of a black and white species pair was made at the near coastal locality of Torra Bay, where white *Onymacris bicolor* and black *O. unguicularis* occur together. At this site on a specific date of census and body temperature measurement, there was essentially no difference between body temperatures of the two species (Fig. 5).

On certain days the situation changes. On the date of the census that provided the data for Figure 5, the absolute activity level for black *O. unguicularis* was nearly three times that of white *O. brincki.* On other dates, without heavy wind, *O. brincki* prevailed.

The lack of a significant difference between the body temperatures when the two species are active together in the area of overlap has several possible explanations. One is that the white beetles are making longer surface forays than the black ones. Preliminary observations indicate this is not the case, as timed individuals generally remained active for 20 minutes or more, and individuals reached thermal equilibrium in less than 5 minutes (Edney, 1971; personal observations). A more likely explanation is based on the fact that the area of sympatry is a particularly windy location. This would result in a convergence in body temperatures for beetles regardless of color. Experimentation with these two species under conditions identical to those noted established that the similarity in body temperature is not the result of identical heat-exchange characteristics. There is also some difference in the choice of preferred habitat on the dunes where these beetles occur together. This difference subjects the white species to more extreme microclimatic conditions.

Where another black and white species pair of similar-sized adesmine tenebrionids coexists, there is a broad area of habitat segregation. In the area of sympatry, the activity of the white species, a white form of the variable species *Onymacris langi* (M. Penrith, personal communication), begins later in the morning and ends earlier in the afternoon than

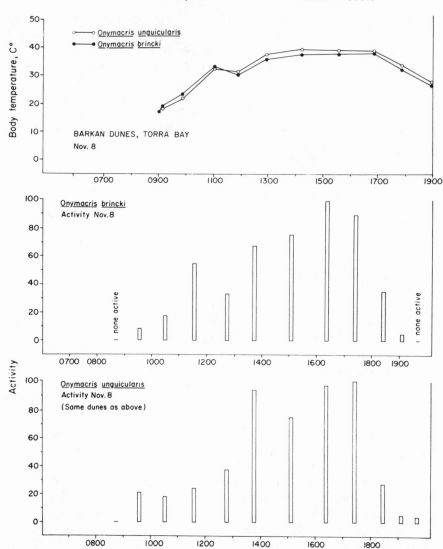

FIGURE 5   Activity cycle and body temperature of the white *Onymacris brincki* and the black *Onymacris unguicularis* beetles on the same Namib Desert dune near the Atlantic coast.

that of its black counterpart, *Physosterna globosa* (Fig. 6). The pattern of the difference in the activity of these two species is as predicted on the basis of coloration. The small extent of the difference was at first surprising to me. However, in retrospect, the small difference in this case and the unmeasurable difference for one day in the case of the *O. brincki–O. unguicularis* white and black species pair is easily

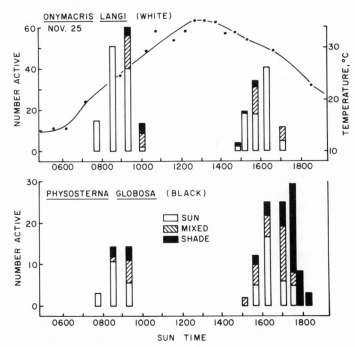

FIGURE 6 Activity cycle of the white *Onymacris langi* and the black *Physosterna globosa* beetles in an area of overlap.

rationalized. If color conveys a significant advantage, but a black and white species coexist in time and space, it follows from competition theory that the advantage of the one over the other must be small and reversible within the range of ambient conditions prevailing at that place.

These data, and this line of analysis, indicate that while activity cycle data in areas of coexistence may be informative, an analysis of the adaptive significance of coloration must depend upon the situation throughout the range of a species. Areas of overlap of black and white species of the Namib Desert Tenebrionidae lie at the geographic edge of the species distribution and at the ecological limits of each species. At the limits of a species distribution, species may be surviving there in spite of rather than because of being fully adapted to the environment.

Body Temperature Measurements

Body temperature optima are suggested here to be the basis of the observed activity patterns. On most occasions, the data gathered to test this prediction were obtained on days when activity cycle data were not obtained, as it was impossible to do both jobs at the same time. Nevertheless, macroclimatic weather data for all dates are available, and from these data it is possible to obtain a general picture of the relationship of body temperature to activity.

Thermal plateau effect. The most convincing evidence for the maxithermy hypothesis is that the insects under consideration here develop and maintain, as a population, body temperatures at the same level, i.e., between 37 and 41°C. Figure 7 provides evidence for three black and one white species of adesmine tenebrionids living in quite different environments. Black *Onymacris rugatipennis* lives in the diverse riverbottom environments, black *Eustolopus octoseriatus* is an inhabitant of the barren, interdune gravel plains, and black *O. laeviceps* lives in diverse environments in the vegetated parts of the Angolan Namib Desert. White *O. langi* lives in an open bunch grass environment where deep sand substrates prevail. In each case, absolute body temperature relationships are essentially the same; there is a period of rising temperature in the morning followed by an interval during which body temperatures plateau between 37 and 41°C. Then, later in the day, as the sources of environmental heat decline, body temperatures fall and surface activity terminates. To emphasize this interpretation, the thermal ascent, plateau, and descent phases of the daily cycle are bracketed by parallel lines in the examples shown in Figure 7. All these data, except those for *O. laeviceps,* were gathered on a single day. To obtain a significant sample, the data for *O. laeviceps* were obtained on four separate dates. One of these dates, August 12, 1967, was relatively cool. The other three dates, August 7, 8, and 9, were about the same and somewhat warmer than August 12.

On all dates for which these data were obtained, the completely exposed high dunes where *O. laeviceps* lives were exposed to extreme thermal conditions during the middle hours of the day. Thus, no individuals were active, and no body temperature data could be obtained for the midday interval. For the same reason only a few of the large-bodied *E. octoseriatus* (Fig. 7) could be located during this same interval.

Holm and Edney (1973) suggest that *O. laeviceps* is crepuscular, based upon their own analysis of the activity rhythm of this species. Since body temperatures reach levels that appear to be maximal (Fig. 7), I conclude that this species is pushed to the ends of the day by the temperature extremes ordinarily encountered in its vegetationless habitat.

Despite spatial differences between the Angolan and South West African parts of the Namib Desert, it is possible to make some tentative comparisons between these regions. Of special interest is that the activity of white *O. candidipennis* tends to be concentrated in the middle hours of the day. On some days in winter, hot east winds prevail. On

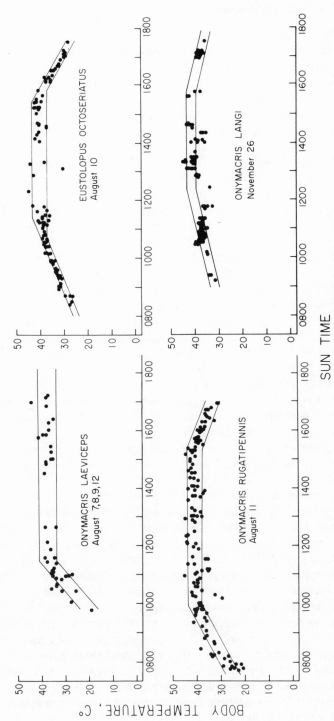

FIGURE 7 Body temperatures of populations of adesmine tenebrionids. Each datum point represents one randomly taken measurement. Measurements were taken as frequently as possible, and no individual was sampled more than once.

85

these occasions, activity may become strongly bimodal. However, the times of peaks in activity are closer to one another than is the case for the black species elsewhere on the desert.

Elsewhere white coloration is relatively uncommon coloration for desert animals. On the Negev Desert, Schmidt-Nielsen et al. (1972) found that the white-shelled snail, *Spincterochila,*was just able to maintain sublethal body temperatures during periods of drought and inactivity.

Color change and thermoregulation. The relative nature of coloration adaptations is well illustrated by the nature of color change phenomena. For desert animals, Norris's (1967) analysis of lizard coloration, notable for its short- and long-term lability, serves well to illustrate the general principle. At low temperatures dark coloration is the rule; and, in general, lizards capable of color change are considerably darker than the background when they first emerge from crevices or underground burrows. For *Dipsosaurus dorsalis,* a well-studied species, body temperatures of 33 to 38°C prevail during this dark phase. In this temperature range these lizards are relatively wary and inactive, and do not range far from cover. With a further rise in body temperature, free-ranging activity begins, the body coloration pales, and the lizard is closely matched to the background in that part of the spectrum visible to the human eye. As the day continues to warm, the lizards blanch still more, becoming gleaming white while maintaining body temperatures in excess of 40°C. They are active into the middle hours of the day, well after their most significant predators have ceased activity. The absolute body temperature levels attained and maintained by lizards are in close agreement with the behavioral actions and thermal responses of beetles. Data suggest an attempt to raise body temperatures below 37°C, to maintain them in the 37 to 41°C range, and to reduce them above this level.

Some tenebrionid beetles are capable of long-term color changes. For example, the long-legged *Stenocara phalangium* emerges in the late fall with a whitish chalky pubescence superimposed on a black substrate; as the winter cools, the pubescence wears away and the animal becomes significantly darker. This same beetle can also accomplish an immediate color change by orienting the more heavily chalked posterior portion of the abdomen toward the sun (Henwood and Hamilton, in press). Henwood has calculated body temperatures for specific conditions in the field for this species, and has found that the ±2°C drop in body temperature accomplished by orientation in this way can be crucial to survival when extreme thermal conditions prevail. In this orientation, the *S. phalangium* beetle has converged to a remarkable degree upon the mountain zebra, *Equus zebra hartmannae,* which also orients the whiter hindquarters toward the sun during extreme hot conditions (Joubert, 1972).

Coloration of homeotherms. Homeotherms may be active at any time during the day. However, under desert conditions, i.e., with limited availability of water, the middle hours of the day are a costly time to be active. Hence activity for most desert homeotherms shifts to the ends of the day. It is probably for this reason that there are few white diurnal desert homeotherms. Hence black coloration may enhance solar radiation absorption, reducing the metabolic cost of homeothermy (Hamilton, 1973).

I conclude that the general significance of black, white, and black and white coloration for diurnal desert animals is that it enables them to be active and to utilize economically such periods of activity. Other colors optimizing alternative adaptions to communication or camouflage must be viewed as compromises with some sacrifice in thermal efficiency.

Other adaptive bases of black and white coloration. The discussion here has emphasized the role of the black and white coloration in thermoregulation. While I imply that a thermal advantage is the leading variable in the evolution of these desert colors, thermal relationships are not the only adaptive basis of these colors. Elsewhere (Hamilton, 1973) I have reviewed alternative hypotheses and identified other consequences of being black in particular. The role of black coloration in Mullerian mimicry is unproved. Melanin does prevent water loss and may be positively advantageous to desert insects in this capacity. A related advantage is the strength of melanin, which in the abrasive desert environment and for long-lived insects in particular may result incidentally in black coloration. These nonthermal advantages of black coloration may account for its occurrence among nocturnal insects. Another possible adaptive significance to black coloration is as a shield against excessive exposure of internal organs to ultraviolet radiation (Porter, 1967). This could explain why all the diverse white Namib Desert Tenebrionidae have black thoraxes and heads. But it cannot account for the black undersides and legs of these beetles. Only a combination of hypotheses would seem to offer a complete explanation of the body colors of these beetles.

SUMMARY

Some desert insects behave so as to maintain a nearly constant body temperature near 38°C when they are active. This level is the same for insects living in diverse environments of significantly different thermal characteristics. No insect maintains a higher level, even though it would be possible to do so, which suggests that this level is the upper limit of thermal existence for these species and that they gain some advantage from maintaining a maximum body temperature level. If this assumption is accepted, the timing of activity, the execution of behavioral actions such as movement in the environmental mosaic, and the relationship of these actions to morphology can be identified as adaptations enhancing the effective maintenance of this level relative to the microclimatic conditions encountered in the environments inhabited by such species.

Acknowledgments

The field aspects of this study were supported by the National Science Foundation (Grant GB-28533). Mr. De la Bat and the staff of Nature Conservation, South West Africa, provided permits and other assistance in the field. I am indebted to M. K. Seely, Director, Namib Desert Research Station, for making available the resources of the Desert Ecological Research Unit at Gobabeb, South West Africa. An earlier draft of the manuscript was read by Eric Edney and Kenneth Henwood, who provided helpful comments. All species observed and reported on in this paper were identified either by the late C. Koch, former Director of the Namib Desert Research Station, or by Mary-Lou Penrith, State Museum, Windhoek.

LITERATURE CITED

Bolwig, N. 1957. Experiments on the regulation of body temperature of certain Tenebrionid beetles. J. Ent. Soc. S. Afr. 20:454–458.

Buxton, P. A. 1924. Heat, moisture, and animal life in deserts. Proc. Roy. Soc. London B96:123–131.

Digby, P. S. B. 1955. Factors affecting the temperature excess of insects in sunshine. J. Exp. Biol. 32:279–298.

Dudley, B. 1964. The effects of temperature and humidity upon certain morphometric and color characters of the Desert Locust (*Schistocerca gregaria* Forskal) reared under controlled conditions. Trans. Roy. Ent. Soc. (London) 116:115–129.

Edney, E. B. 1967. Water balance in desert arthropods. Science 156:1059–1066.

———. 1971. The body temperature of tenebrionid beetles in the Namib Desert of Southern Africa. J. Exp. Biol. 55:253–272.

Ellis, P. E., and D. B. Carlisle. 1961. The prothoracic gland and colour change in locusts. Nature 190:368–369.

Gates, D. M. 1974. Animal coloration. Review of "Life's color code" by W. J. Hamilton III. Bioscience 24:120.

Hadley, N. F. 1970. Micrometeorology and energy exchange in two desert arthropods. Ecology 51:434–444.

Hamilton, W. J., III. 1971. Competition and thermoregulatory behavior of the Namib Desert tenebrionid beetle genus *Cardiosis*. Ecology 52:810–822.

———. 1973. Life's color code. McGraw-Hill, New York. 238 pp.

Hellmich, W. 1933. Die biogeographischev Grundlagen Chiles. Zool. J., Abt. Syst. 64:165–226.

Henwood, K. In press. A field tested thermoregulation model for two diurnal Namib Desert tenebrionid beetles. Ecology.

———, and W. J. Hamilton. In preparation. Morphological color change in a desert beetle and its adaptive significance.

Holm, E., and E. B. Edney. 1973. Daily activity of Namib Desert arthropods in relation to climate. Ecology 54:45–56.

Joubert, E. 1972. Activity patterns shown by Hartmann Zebra *Equus zebra hartmannae* in South West Africa with reference to climatic factors. Madoqua, Ser. 1, 5:33–52.

Key, K. H. L., and M. F. Day. 1954a. A temperature-controlled physiological colour response in the grasshopper *Koscuiscola tristis* Sjost (Orthoptera: Acrididae). Aust. J. Zool. 2:309–339.

———, and M. F. Day. 1954b. The physiological mechanism of colour change in the grasshopper *Koscuiscola tristis* Sjost. (Orthoptera: Acrididae). Aus. J. Zool. 2:340–363.

Koch, C. 1961. Some aspects of abundant life in the vegetationless sand of the Namib Desert dunes. Publ. Namib Desert Res. Stn. 1:8–34, plus 10 plates.

———, 1962. The Tenebrionidae of southern Africa: XXXI. Comprehensive notes on the tenebrionide fauna of the Namib Desert. Publ. Namib Desert Res. Sta. 5:61–106.

Nolte, D. J. 1964. Chiasma frequency and gregarization in locusts. Nature 204:1110–1111.

Norris, K. S. 1967. Color adaptation in desert reptiles and its thermal relationships. In W. W. Milstead (ed.), Lizard ecology, a symposium. University of Missouri Press, Columbia, Mo.

Pepper, J. H., and E. Hastings. 1952. The effects of solar radiation on grasshopper temperature and activities. Ecology 33:96–103.

Porter, W. P. 1967. Solar radiation through the living body walls of vertebrates with emphasis on desert reptiles. Ecol. Monogr. 33:273–296.

———, and D. M. Gates. 1969. Thermodynamic equilibria of animals with environment. Ecol. Monogr. 39:227–244.

Schmidt-Nielsen, K., C. R. Taylor, and A. Shkolnik. 1972. Desert snails: problems of survival. Symp. Zool. Soc. London 31:1–13.

THERMOREGULATION AND FLIGHT
ENERGETICS OF DESERT INSECTS

Bernd Heinrich

Department of Entomological Sciences,
University of California, Berkeley, California

Abstract

Insects in deserts face the potential problems of overheating and desiccation. Both are magnified during flight, particularly in large insects. Obligatory endothermy as a result of flight metabolism raises thoracic temperature more than 30°C above ambient in some insects weighing ½ g or more. The high metabolic rate, compounded with endothermy, results in high rates of water loss from respiratory surfaces; insects can desiccate in flight even in an atmosphere saturated with water vapor. However, water is conserved by expiring through, or in and out, the relatively cool abdomen rather than through the hot thorax. Respiratory water loss, and the evaporative cooling resulting from it, are computed for three large insects found in deserts: the desert locust, bees, and sphinx moths. Evaporation removes 8 to 9 percent of the total heat produced in the thorax, but the difference in thoracic temperature resulting from evaporative cooling during flight in dry air and in air fully saturated with water vapor is relatively insignificant. Active cooling by the circulatory system is observed, but this mechanism is only effective in large insects, which generate a marked difference between thoracic and ambient temperatures. Physiological limitations place great importance on activity patterns that ameliorate temperatures and water stress.

INTRODUCTION

The physical challenges of the desert environment to organisms are primarily high temperatures and low humidities. Insects are preadapted to these potential problems because of their structure and physiology. Small size allows them to utilize microenvironments for both thermal and water balance (Cloudsley-Thompson, 1964; Edney, 1967), and the excretory, respiratory, and integumentary systems can be adapted for water retention.

Some ground-dwelling insects, like the desert cockroach *Arenivaga* (Edney, 1967) and tenebrionid beetles (Ahearn and Hadley, 1969; Ahearn, 1970), may owe their success in desert habitats to specific features allowing them to cope with high temperatures and little water. However, other insects living in deserts do not appear to be morphologically or physiologically uniquely adapted in any obvious way. These insects include dragonflies, numerous bees (including the imported honeybee *Apis mellifera*), and lepidoptera such as sphinx moths *Hyles lineata* and *Manduca sexta* (all of which are also found in a wide range of other habitats).

Thermal and water balance pose potential problems particularly during flight where they are interlinked. Insects generate large amounts of heat as a result of flight metabolism, which can become a liability at high ambient temperatures. The large amounts of air ventilating thoracic muscles during flight result in water loss, but this water loss results in a small amount of evaporative cooling. I shall here examine interrelationships among flight energetics, water loss, and body temperature, and discuss their possible implications to insects living in deserts.

BODY TEMPERATURES IN FLIGHT

The major portion of the considerable energy expended by insects during flight is degraded to heat (Weis-Fogh, 1972). However, in small insects (< 100 mg) the rate of convective heat loss during flight is sufficiently rapid so that relatively little of this heat is retained in the muscles (Digby, 1955), and thoracic temperature (T_{Th}) probably does not increase more than 1°C above ambient temperature (T_A). In contrast, moths (heavily insulated with pile) weighing only 1 g may generate a T_{Th} of 30°C above T_A in flight (Bartholomew and Heinrich, 1973). Abdominal temperature tends to remain relatively close to ambient temperature.

The effect of insulation on the temperature excess $(T_{Th} - T_A)$ was examined experimentally by Church (1960b). He observed that denuded dead moths and bumblebees (heated internally with a resistor) had a temperature excess related nearly directly to the diameter of their pterothorax (Fig. 1). The temperature excess of sphinx moths covered with pile was approximately double that when they are naked; in bumblebees, insulation accounted for approximately 65 percent of the temperature excess.

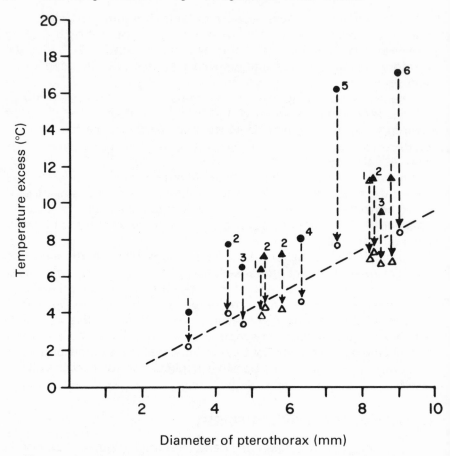

FIGURE 1 Effect of pile and diameter of pterothorax on the temperature excess ($T_{Th} - T_A$) in a 300-cm/s wind (adapted from Church, 1960b). Circles = moths, triangles = bumblebees; filled (dark) symbols = before depilation, open = after depilation. Regression line refers to denuded insects only. Animals had been recently killed and were heated with a resistor implanted in the thorax. Heat input was proportional to the volume of the pterothorax. (A 9-mm-diameter thorax received an input of 1.94 cal/min and the 3.2-mm-diameter thorax was heated with 0.08 cal/min.)

Moths

1 = *Agrotis puta* (Hueb.)
2 = *Malacosoma neustria* Linn.
3 = *Plusia gamma* Linn.
4 = *Triphaena pronuba* Kubb.
5 = *Laothoe populi* Linn.
6 = *Sphinx ligustri* Linn.

Bumblebees

1 = *Bombus terrestris* Linn.
2 = *B. lapidarius* Linn.
3 = *Psithyrus vestalis* Foucr.

Church used heat inputs considerably less than those actually generated by the insects in flight. For example, sphinx moths (*Manduca sexta*) and bumblebees (*Bombus edwardsii*) with a thoracic diameter of about 10 and 6 mm, respectively, expend approximately 7.0 and 2.3 cal/min in flight (Heinrich, 1971a, 1974b), of which more than 80 percent is degraded to heat. Church's heat inputs for animals of this size were 2.7 and 0.5 cal. Therefore, the "passive" heating in live animals (due to flight metabolism) should considerably exceed his observed temperatures from heating with a thermode (Fig. 1). Despite the potentially large temperature excess resulting from the high rate of heat production in the thorax during flight, the sphinx moth *M. sexta* from the Mojave desert is heavily insulated on the thorax, although the abdomen is lightly insulated (Heinrich, 1971a).

Available data suggest no obvious correlations of metabolic rate to body weight of insects in flight (Fig. 2). However, the rate of passive heat loss *is* related to body size, and, thus, the limited capacity for endothermy in flight of small insects is obvious. The small thoracic temperature excess of small insects in flight† greatly limits their scope for thermoregulation. Nevertheless, a comparative study of numerous species of moths reveals that the effect of body size on thoracic temperature in these insects is no longer apparent at body weights greater than approximately ½ g (Fig. 3). Even at T_A of 16°C or less, T_{Th} can approach 45°C. It appears that mechanisms of thermoregulation that prevent overheating in larger insects during continuous flight become operative.

Honeybees are medium-sized insects ($\approx$ 60 to 100 mg), yet they maintain a temperature excess nearly 20°C above T_A during continuous flight (B. Feldmeth, personal communication). Thoracic temperatures of honeybees in excess of 44.5°C have not been measured, but honeybees enter and leave their hive even at T_A near 43°C (D. B. Dill, personal communication). [Bumblebees do not remain in continuous flight in a temperature-controlled room at $T_A > 35°C$ (Heinrich, 1975)]. These ob-

---

† The basic equation describing heat production per body weight (*Hp*) and body temperature excess (ΔT), assuming weight (*W*) is equivalent to volume (*V*) is $HpV = WS\ dT_B/dt\ CA(\Delta T)$ (see page 96). Since area (*A*) is proportional to $W^{2/3}$ and since there is no change in T_B during steady-state flight, the equation becomes $HpW = CW^{2/3}(\Delta T)$, in which $\Delta T = Hp/CW^{1/3}$. From this we can get an approximation of ΔT_1, the temperature excess of a small (*W* = 2.0 mg) insect such as *Drosophila* in flight, by comparing it with the known ΔT_2 (6°C) of an insect such as *Schistocerca,* which has a similar metabolic rate during flight as does *Drosophila* and which weighs 2000 mg. Assuming the two insects have similar conductance (*C*), the ratio $\Delta T_1/\Delta T_2 = (W_1/W_2)^{1/3}$ can be used to solve ΔT_1. In this case, for *Drosophila* it is 0.6°C.

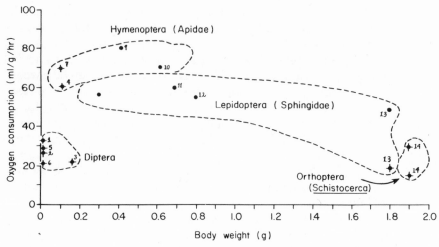

FIGURE 2 Rate of oxygen consumption during flight in insects of various sizes.
1 = *Drosophila,* calculated from sugar consumption (Hocking, 1953); 2 = *Simulium venustum (ibid.);* 3 = *Tabanus affinus (ibid.);* 4 = *Apis mellifera (ibid.);* 5 = *Drosophila americana* (Chadwick, 1947); 6 = *Drosophila repleta* (Chadwick and Gilmour, 1940); 7 = *Apis mellifera* (Bastian and Esch, 1970); 8 = *Mimas tiliae* (Zebe, 1954); 9 = *Bombus edwardsii* (Heinrich, 1975); 10 = *B. vosnesenskii (ibid.);* 11 = *Deilephila elpenor* (Heinrich and Casey, 1973); 12 = *Hyles euphorbia (ibid.);* 13 = *Manduca sexta* (Heinrich, 1971a); 14 = *Schistocerca gregaria* (Krogh and Weis-Fogh, 1951; Weis-Fogh, 1964.) • = free flight, ✚ = tethered.

servations suggest that honeybees either have a very efficient cooling mechanism during flight and/or their flight is not "continuous" at high T_A.

Bees fly at 6.5 to 7.5 m/s (Wenner, 1963) and could thus fly nearly ½ k in only 1 min, which probably is sufficient time for their T_{Th} to equilibrate (as occurs, by definition, in "continuous" flight). Honeybees maintain their hive temperature well below 40°C even at T_A near 48°C (Lensky, 1964) by regurgitating liquid and sprinkling water onto combs. Vigorous fanning of this surface results in evaporative cooling (Lindauer, 1954). Perhaps bees use the hive as a site to unload heat, analogous to the desert ground squirrel's use of its burrow (Bartholomew and Hudson, 1961). If so, it could be predicted that even though the need for and the number of flights for water increase with increasing T_A, the *distance* (time) the bees are able to travel decreases. To my knowledge these aspects relating to the honeybees' survival in the desert have not been investigated.

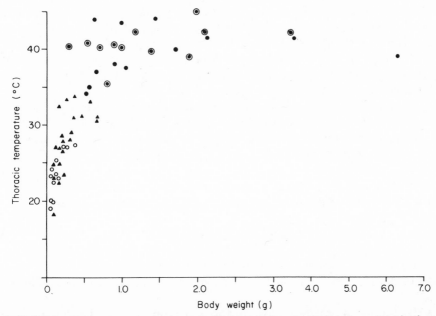

FIGURE 3 Thoracic temperatures in flight as a function of body weight in three families of moths. ▲ = Arctiidae; o = Ctenuchidae; ◉ = Sphingidae at 20°C; ● = Sphingidae at 7°C and 15 to 17°C. Each point represents the mean of a species. (Data from Bartholomew and Heinrich, 1973, and Heinrich and Casey, 1973.)

ENERGETIC COST OF WARM-UP AND FLIGHT

Unless actively cooled, flight muscles of larger insects unavoidably heat up to 30°C or more above ambient temperature during continuous flight. It is therefore not surprising that they are adapted to operate at much higher temperatures than ambient. Thus, to initiate flight, such insects must first warm up.

Numerous diurnal insects warm up by basking (see Fraenkel, 1930; Vielmetter, 1954; Watt, 1969), directly utilizing solar radiation. In many other insects, including nocturnal moths, warm-up involves endothermy that results from shivering by flight musculature. When these insects are active in the absence of sunshine, the energetic cost of a flight includes also that of warm-up.

Physiological mechanisms of endothermic warm-up in insects have been investigated on numerous occasions (Krogh and Zeuthen, 1941; Kammer, 1968), but little is known about the energetic costs involved. The metabolic rate and T_{Th} change very rapidly throughout a warm-up bout. Although it is difficult to measure metabolism directly, T_{Th} can easily be measured. From this T_{Th} and the passive cooling curves, the metabolic rate can be calculated (Heath and Josephson, 1970; Heinrich and Bartholomew, 1971), provided heat loss remains passive. The metabolic rate of the sphinx moth *Manduca sexta* during warm-up at different T_{Th} and T_A has been computed by this method (Heinrich and Bartholomew, 1971).

Calculations of energy expenditure during warm-up are simplified by the fact that moths leak little heat into the abdomen during warm-up. Total heat production (heat storage + heat loss) can be calculated as follows from thoracic weight ($W = 0.5$ g), its specific heat ($S = 0.8$ cal/g °C), rate of passive cooling (e.g., 3.9°C/min at $T_{Th} - T_A = 25°C$ in a dead *M. sexta*), and the warm-up curves (Heinrich and Bartholomew, 1971). The rate of passive heat loss is expressed by the equation

$$WS \frac{dTh}{dt} = AC(T_{Th} - T_A) \tag{1}$$

where A = surface area and C = thermal conductance of the surface. Using data on cooling rates of dead moths in still air yields 0.0624 for AC.

The rate of heat production during warm-up (dHp/dt) equals the rate of heat storage ($WS\ dT_{Th/dt}$) plus the rate of heat loss [$AC(T_{TH} - T_A)$] (Heath and Josephson, 1970). Integrating both sides of the equation gives the expression for the amount of heat produced during the entire warm-up at a given T_A:

$$Hp(T_A) = WS(T_f - T_A) - ACT \cdot t_f + AC \int_0^{t_f} T_{Th}\ dt \tag{2}$$

where T_f = final T_{Th} and t_f = time until T_f. Solving for H_P at different T_A on the basis of the observed warm-up curves shows that it costs a moth 4 cal to warm up at 30°C, and at least 15 cal at 15°C (Fig. 4). Small amplitude vibrations of the wings and respiration during warm-up cause additional heat loss due to convection and possibly evaporation (about 0.08 cal/min/°C; Heinrich and Bartholomew, 1971). Adding this heat loss, for example, to the cost of warm-up at 15°C yields 25 cal, the "total" energy cost of warm-up in a wind-still environment.

It is of obvious energetic advantage to warm up as rapidly as possible. Moths appear to do this insofar as their body temperature allows it. Their rate of heat production is related to muscle temperature, and thus they warm up much more rapidly at high than at low T_A. Consequently they have a lower *overall* energy expenditure for a given warm-up at high in comparison to low ambient temperatures.

The metabolic cost of flight in insects varies greatly (Fig. 2). Some factors affecting this cost are discussed by Heinrich (1974). Variables include frequency of muscle contractions (wingbeats) and wing load-

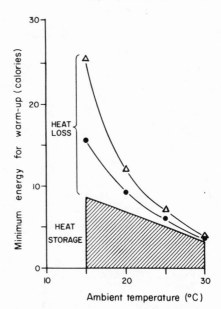

FIGURE 4 The calculated total energy expenditure by a sphinx moth *Manduca sexta* (1.5 g) from the beginning of warm-up until the initiation of flight as a function of ambient temperature (no wind). • = still air; △ = includes heat loss due to forced convection resulting from wing vibrations. Cross-hatching = portion of energy expenditure appearing as stored heat. (Data on which these calculations are based are from Heinrich and Bartholomew, 1971.)

ing. Tethered insects tend to have considerably lower metabolic costs in flight than insects in free flight. High work output during flight demands high thoracic temperatures, and results in high rates of heat production. A moth weighing 2 g, for example, produces almost 8 cal of heat in the thorax per minute of free flight.

WATER LOSS AND THORACIC TEMPERATURE DURING FLIGHT

To maintain their high metabolic rates during flight, the muscles in the thorax must be continuously supplied with large amounts of air and heat must be removed. Oxygen is absorbed through moist respiratory surfaces; hence ventilation is accompanied by evaporative water loss. The higher the temperature of the surface from which evaporation occurs, the more water will be lost. High metabolic rates and high thoracic temperatures thus imply concomitantly high water costs. Water loss is also affected by the direction that air moves through the tracheal system and the substrate metabolized during flight.

I shall now discuss comparative aspects of water balance in three taxonomically, physiologically, and behaviorally different insects found in deserts: a grasshopper (the desert locust), a bee, and a sphinx moth. To simplify the comparisons, I shall restrict discussion to flight in dry (0 percent r.h.) and fully saturated (100 percent r.h.) air, and ignore cuticular and other water loss not directly associated with ventilation and thermoregulation. I shall also assume that inspired air becomes saturated with water vapor. During average horizontal flight, thoracic ventilation in the desert locust *Schistocerca gregaria* is about 320 ml air·g^{-1}·h^{-1}, and abdominal ventilation is 180 ml·g^{-1}·h^{-1} (Fig. 5). Of the latter, 30 ml·g^{-1}·h^{-1} are moved unidirectionally from thorax to abdomen (Weis-Fogh, 1967). Combining these data with the known endothermy and substrate metabolized during flight, it is possible to estimate water balance.

Weis-Fogh (1967), who has discussed water balance of *Schistocerca* in terms of flight energetics, states that flight metabolism of these animals yields approximately 8 mg of water ·g^{-1}·h^{-1} (Table 1). Locusts are endothermic only to a limited extent, and fly at relatively high ambient temperatures (Weis-Fogh, 1964). During flight at 30°C with a T_{Th} of 37°C, they should lose approximately 9 mg of water·g^{-1}·h^{-1} in dry air. At 100 percent r.h. the rate of water loss is quite low in comparison to the rate of metabolic water production, resulting in a net gain in water (Table 1).

It was assumed for the calculations that expired air is saturated with water vapor at the temperature of the thorax or abdomen where it is expired. The greater the endothermy of the body at this exit point, the more water a given volume of expired air would remove. When fully saturated with water vapor, the 30 ml of air·g^{-1}·h^{-1} moved unidirectionally from thorax to abdomen would contain 0.90 mg of water at

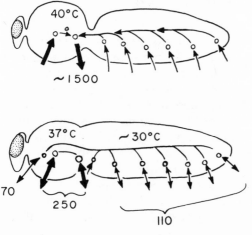

20°C BEE

~1500

37°C   ~30°C

70

250

110

30°C "LOCUST"

FIGURE 5 Approximate body temperatures during flight and diagrammatic indication of directions and approximate volume (ml·g^{-1}·h^{-1}) of air through the tracheal system of flying honeybees, *Apis mellifera* (Bailey, 1954), and desert locusts, *Schistocerca gregaria* (based on data from Weis-Fogh, 1967, and Miller, 1960).

TABLE 1 Calculations of respiratory water loss and percentage of heat removed by evaporation for three insects in flight

	Oxygen uptake $(ml \cdot g^{-1} \cdot h^{-1})$	Substrate	Metabolic water $(mg \cdot g^{-1} \cdot h^{-1})$	Ventilation $(ml \cdot g^{-1} \cdot h^{-1})$	T_{Th} (°C)	T_A (°C)	Respiratory water loss $(mg \cdot g^{-1} \cdot h^{-1})$ 0% r.h.	100% r.h.	Water balance $(mg \cdot g^{-1} \cdot h^{-1})$ 0% r.h.	100% r.h.	Heat removal by evaporation (%) 0% r.h.	100% r.h.
Schistocerca	15[a]	Lipid	7.94	320 (thorax) 110 (abdomen)	37 (30)	30	13.9 3.30	4.26 0	−9.27	+3.68	1.1	0.3
Sphinx moth	50	Lipid	26.5	1070[b]	40	20	53.9	35.6	−27.5	−9.14	13	8.8
Bee	70	Carbohydrate	48.7	1490[b]	40	20	75.1	49.6	−26.5	−0.94	12	8.2

[a] At 100 percent lift.
[b] Assumed same extraction efficiency of oxygen from air as in Schistocerca.

30°C, and 1.30 mg at 37°C. *Schistocerca* may thus save 0.40 mg of water·g^{-1}·h^{-1} during flight at 30°C by expiring the unidirectionally flowing air through the abdomen rather than through the thorax. This assumes, however, that water deposited by condensation in the tracheal trunks reaches the tracheoles (perhaps by capillary action) from which it could be absorbed. The remaining 80 ml of air·g^{-1}·h^{-1} of abdominal ventilation (Fig. 5) is probably moved tidally in and out of the abdomen (Weis-Fogh, personal communication). An additional 1.07 mg of water·g^{-1}·h^{-1} would be saved by this arrangement, rather than by expiring the air through the thorax, as in the honeybee (Fig. 5).

A bee metabolizes carbohydrate in flight. Having a respiratory rate near 70 ml of O_2·g^{-1}·h^{-1} (Fig. 2), it must ventilate its flight muscles with approximately 1500 ml of air·g^{-1}·h^{-1} (Table 1). Unlike the desert locust, the bee inspires through the abdomen (Bailey, 1954) so that all of its expired air is presumably close to T_{Th}. Since the bee is more endothermic than the locust and since all air is expired by way of the thorax, it loses considerably more water during flight than does the locust. At a T_A of 20°C and a temperature excess of 20°C, the bee would lose water at a rate of nearly 1 mg·g^{-1}·h^{-1} in fully saturated air. In dry air the bee would desiccate fairly rapidly, losing water at the rate of 2.65 percent body weight per hour (Table 1).

Moths metabolize lipid in flight (Kozhantshikov, 1938; Beenakkers, 1969). Although 1 mg of "typical" fat provides 1.07 mg of metabolic water and 1 mg of "mixed" carbohydrate provides only 0.556 mg, 2.375 mg of carbohydrate must be combusted for every milligram of lipid to provide the same number of calories in aerobic respiration. Furthermore, slightly more oxygen is needed to provide a given number of calories through aerobic respiration of lipid (0.213 ml/cal) than through aerobic respiration of carbohydrate (0.200 ml/cal). As a result, more water is lost on an isocaloric intake of lipid than carbohydrate. Carhohydrate should be a better diet for water balance under desiccation conditions than lipid, but its inferior weight economy for a substrate during flight may be of overall energetic disadvantage.

On the basis of substrate during flight, temperature excess, and the observation that there is little flow of air between thorax and abdomen (Fig. 6, Table 2), indicating that most of the air is inspired as well as expired from the thorax, moths should desiccate more rapidly than bees during flight in dry air. Their rate of desiccation during flight at 20°C and 100 percent r.h. should be nearly the same as that of the locust flying at much higher temperature in *dry* air (30°C and 0 percent r.h.). Considering body temperature, metabolic rate, and associated water loss, the desert locust appears to be better adapted to withstand desiccation during flight than the bee or moth.

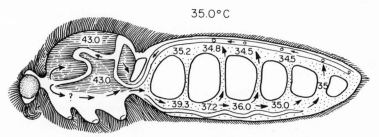

35.0°C

FIGURE 6 Diagrammatic representation of the circulatory system of a sphinx moth, and the body temperatures observed during flight at 35°C. (Adapted from Heinrich, 1971b.)

EVAPORATIVE COOLING AS A RESULT OF VENTILATION IN FLIGHT

Water evaporated from moist respiratory surfaces results in cooling. Since each milligram of evaporated water removes 0.38 cal, and because the total number of calories produced in the thorax can be computed from the respiratory rate, the magnitude of the cooling effect can be readily determined. As shown experimentally (Church, 1960a), the cooling effect is not great when it is assumed that no evaporative cooling occurs at 100 percent r.h. In *Schistocerca* it amounts to 1°C or less even under very desiccating conditions. However, calculations suggest that at a temperature excess of 20°C bees and moths should lose 8 or 9 percent of their metabolic heat by evaporative water loss in a saturated atmosphere (Table 1), and heat loss would increase to 12 to 13 percent in dry air. Thus, the difference between the amount of heat that is lost at 0 and 100 percent r.h. amounts to less than 1°C *difference* between the possible temperature excess of the thorax during flight at 0 and 100 percent r.h. This is what has been measured experimentally, but, for reasons given previously, it would be wrong to presume that the evaporative cooling is only 1°C. The calculations suggest that the pass-

TABLE 2 Movement of air through the sphinx moth *Manduca sexta* in tethered flight at 20 to 22°C (Heinrich, unpublished)[a]

	Duration of flight $\bar{x}$ (min)	N	Ml of air·g^{-1}·h^{-1} Range	$\bar{x}$
Into abdomen	11.3	5	0.15–3.30	1.52
No movement	10.0	6	—	0.00
Out of abdomen	8.5	4	0.10–0.50	0.23

[a] Overall $\bar{x}$ = 0.45 ml of air·g^{-1}·h^{-1} into abdomen.

ive cooling amounts to several degrees Celsius both in dry and saturated air.

ACTIVE COOLING

The dissipation of excess heat in vertebrates involves considerable expenditure of both energy and water. As shown, overheating of some insects in flight is also a potential problem. Sphinx moths during hovering flight do not vary their metabolic rate as a function of ambient temperature (Heinrich, 1971a; Heinrich and Casey, 1973), yet the temperature excess, which is 24°C at T_A of 22°C or less, is reduced to 8°C at T_A of 30°C or more. The reduction of the temperature excess clearly involves active cooling. The mechanism for this cooling of flight muscles in the sphinx moth *M. sexta* involves the transport of heat by blood (Fig. 6) from the thoracic muscles to the abdomen (Heinrich, 1971b). In effect, the mechanism requires negligible energy expenditure and does not make demands on the insect's water supplies. However, the effectiveness of the cooling mechanism is limited to relatively large temperature excesses. As T_A increases and T_{Th} reaches the upper physiological limit, the mechanism becomes increasingly inefficient. On the basis of theoretical considerations, neither evaporative cooling nor the described cooling mechanism would suffice to achieve thermal balance in large insects during continuous flight at the high T_A commonly observed in the desert during daytime.

ACTIVITY PATTERNS AND THERMAL–WATER BALANCE

The preceding considerations have illustrated potential difficulties large insects should encounter during continuous flight at high ambient temperatures. Nevertheless, large highly endothermic insects like sphinx moths and bees, which must lose relatively large amounts of water during flight, are not rare in the deserts and tropics (Heinrich, 1972). Their heat and water balance, however, cannot be considered in isolation from their patterns of activity.

Of the several large insects discussed here only the desert locust is known to engage in prolonged flight in the daytime when ambient temperatures are high and relative humidity is low. The locust's relatively low metabolic rate in flight, together with nervous control of the spiracles (Miller, 1960; Loveridge, 1968), appears to make it physiologically "desert-adapted." (Little is known about spiracular control by bees and moths in flight.) On the other hand, large diurnal cicadas native to the desert overheat after several seconds of flight at temperatures at which they are normally "active" (Heath and Wilkin, 1970). However, at high T_A cicadas remain in the shade of vegetation. There they sing and feed on plant sap while remaining relatively stationary. In contrast, sphinx moths feed while hovering, and would thus be incapable of frequent diurnal activity in the desert owing to overheating. By deferring activity to night, when ambient temperatures are lower and relative humidity is higher, both water and temperature problems are markedly ameliorated. The sphinx moth *Hyles lineata* is active both diurnally and nocturnally. I have observed the moths in the Anza Borrego desert only at night, but in the Sierras moths were seen foraging in the daytime.

An alternative that would prevent overheating and reduce water loss from the spiracles during activity (Ahearn and Hadley, 1969; Ahearn, 1970) is a reduction in the need for respiratory gas exchange. This is virtually assured in a flightless insect. Perhaps the tenebrionid beetles owe part of their conspicuous success in the deserts (Edney, 1971; Hamilton, 1971) to this way of life.

Acknowledgments

I thank George F. Oster and Eric B. Edney for helpful criticisms. Supported by NSF Grant GB-31542.

LITERATURE CITED

Ahearn, G. A. 1970. The control of water loss in desert Tenebrionid beetles. J. Exp. Biol. 53:573–595.

——, and N. F. Hadley. 1969. The effects of temperature and humidity on water loss in two desert Tenebrionid beetles, *Eleodes armata* and *Cryptoglossa verruscosa.* Comp. Biochem. Physiol. 30:739–749.

Bailey, L. 1954. The respiratory currents in the tracheal system of the adult honey-bee. J. Exp. Biol. 31:589–593.

Bartholomew, G. A., and B. Heinrich. 1973. A field study of flight temperatures in moths in relation to body weight and wing loading. J. Exp. Biol. 58:123–135.

——, and J. W. Hudson. 1961. Desert ground squirrels. Sci. Amer. 205:107–116.

Bastian, J., and H. Esch. 1970. The nervous control of the indirect flight muscles of the honey bee. Z. vergl. Physiol. 67:307–324.

Beenakkers, A. M. T. 1969. Carbohydrate and fat as a fuel for insect flight. A comparative study. J. Insect Physiol. 14:353–361.

Chadwick, L. E. 1947. The respiratory quotient of *Drosophila* in flight. Biol. Bull. 93:229–239.

——, and D. Gilmour. 1940. Respiration during flight in *Drosophila repleta* Wollaston: the oxygen consumption considered in relation to wing-rate. Physiol. Zool. 13:398–410.

Church, N. S. 1960a. Heat loss and the body temperatures of flying insects. I. Heat loss by evaporation of water from the body. J. Exp. Biol. 37:171–184.

————. 1960b. Heat loss and the body temperatures of flying insects. II. Heat conduction within the body and its loss by radiation and convection. J. Exp. Biol. 37:187–212.

Cloudsley-Thompson, J. L. 1964. Terrestrial animals in dry heat: arthropods. In D. B. Dill, E. F. Adolph, and C. G. Wilber (eds.), Handbook of physiology, pp. 451–465. American Physiological Society, Washington, D.C.

Digby, P. S. B. 1955. Factors affecting the temperature excess of insects in sunshine. J. Exp. Biol. 32:279–298.

Edney, E. B. 1967. Water balance in desert arthropods. Science 156:1059–1066.

————. 1971. The body temperatures of Tenebrionid beetles in the Namib desert of Southern Africa. J. Exp. Biol. 55:253–272.

Fraenkel, G. 1930. Die Orientierung von *Schistocerca gregaria* zu strahlender Wärme. Z. vergl. Physiol. 13:300–313.

Hamilton, W. J., III. 1971. Competition and thermoregulatory behavior of the Namib desert Tenebrionid beetle Genus *Cardiosis*. Ecology 52:810–822.

Heath, J. E., and R. K. Josephson. 1970. Body temperature and singing in the Katydid, *Neoconocephalus robustus* (Orthoptera, Tettigoniidae). Biol. Bull. 138:272–285.

————, and P. J. Wilkin. 1970. Temperature responses of the desert cicada *Diceroprocta apache* (Homoptera, Cicadidae). Physiol. Zool. 43:145–154.

Heinrich, B. 1971a. Temperature regulation of the sphinx moth, *Manduca sexta*. I. Flight energetics and body temperature during free and tethered flight. J. Exp. Biol. 54:141–152.

————. 1971b. Temperature regulation of the sphinx moth, *Manduca sexta*. II. Regulation of heat loss by control of blood circulation. J. Exp. Biol. 54:153–166.

————. 1972. Thoracic temperatures of butterflies in the field near the equator. Comp. Biochem. Physiol. 43A:459–467.

————. 1974. Thermoregulation in insects. Science 185:747–756.

————. 1975. Thermoregulation in bumblebees. II. Energetics of warm-up and free flight. J. Comp. Physiol. 96:155–166.

————, and G. A. Bartholomew. 1971. An analysis of pre-flight warm-up in the sphinx moth, *Manduca sexta*. J. Exp. Biol. 55:223–239.

————, and T. M. Casey. 1973. Metabolic rate and endothermy in sphinx moths. J. Comp. Physiol. 82:195–206.

Hocking, B. 1953. The intrinsic range and speed of flight of insects. Trans. Roy. Ent. Soc., London 104:223–345.

Kammer, A. E. 1968. Motor pattern during flight and warm-up in Lepidoptera. J. Exp. Biol. 48:89–109.

Kozhantshikov, I. W. 1938. Carbohydrate and fat metabolism in adult Lepidoptera. Bull. Ent. Res. 29:103–115.

Krogh, A., and T. Weis-Fogh. 1951. The respiratory exchange of the desert locust *(Schistocerca gregaria)* before, during and after flight. J. Exp. Biol. 28:344–357.

————, and E. Zeuthen. 1941. The mechanism of flight preparation in some insects. J. Exp. Biol. 18:1–10.

Lensky, Y. 1964. Comportement d'une colonie d'abeilles a des temperatures extremes. J. Insect Physiol. 10:1–12.

Lindauer, M. 1954. Temperaturegulierung und Wasserhaushalt in Bienenstaat. Z. Vergl. Physiol. 36:391–431.

Loveridge, J. P. 1968. The control of water loss in Locusta migratoria migratorioides R. & F. J. Exp. Biol. 49:15–29.

Miller, P. L. 1960. Respiration in the desert locust. III. Ventilation and the thoracic spiracles during flight. J. Exp. Biol. 37:264–278.

Vielmetter, W. 1954. Die Temperaturregulation des Kaisermantels in der Sonnenstrahlung. Naturwiss. 41:535–536.

Watt, W. B. 1969. Adaptive significance of pigment polymorphisms in Colias butterflies, II. Thermoregulation and photoperiodically controlled melanin variation in Colias eurytheme. Proc. Nat. Acad. Sci. (U.S.) 63:767–774.

Weis-Fogh, T. 1964. Biology and physics of locust flight VIII. Lift and metabolic rate of flying locusts. J. Exp. Biol. 41:257–271.

———. 1967. Respiration and tracheal ventilation in locusts and other flying insects. J. Exp. Biol. 47:561–587.

———. 1972. Energetics of hovering flight in hummingbirds and Drosophila. J. Exp. Biol. 56:79–104.

Wenner, A. M. 1963. The flight speed of honeybees: a quantitative approach. J. Apic. Res. 2:25–32.

Zebe, E. 1954. Über den Stoffwechsel der Lepidopteren. Z. Vergl. Physiol. 36:290–317.

NITROGEN EXCRETION IN ARID-ADAPTED AMPHIBIANS

Lon L. McClanahan

Department of Biology, California State University at Fullerton, Fullerton, California

Abstract

Patterns of nitrogen excretion were surveyed in arid-adapted amphibians. It was concluded that the lability of nitrogen excretion in amphibians is as great as that found in many reptiles. Ureotelism has been exploited by many fossorial desert amphibians as a device to detoxify ammonia. If urea is stored in the body fluids, it will lower body water potentials so that water loss to the soil is minimized. Uricotelism has been adopted by many arboreal anurans but has become especially effective as a water conservation device in those species that have evolved mechanisms to limit evaporative water loss. The capability of many larval anurans to excrete nitrogen as urea is reviewed.

INTRODUCTION

Research on nitrogen excretion in amphibians has led to broad generalizations based on observations from a few species. Within the last few years amphibians from differing habitats have been investigated and found not to fit these generalizations. It is not my intent to make an exhaustive review of the literature dealing with this subject since other excellent reviews already exist (Balinski, 1970; Campbell and Goldstein, 1972). However, I would like to review some of the more recent studies that demonstrate the lability of excretory patterns in arid-adapted amphibians, particularly anurans, since there is little information available on arid-adapted urodeles.

Based on its widespread occurrence in amphibians, ureotelism appears to have evolved early in the history of this group. The necessity to detoxify ammonia and to alter body water potentials to achieve osmotic equilibrium were most likely strong selective determinants in that direction. When many aquatic anurans are deprived of free water by dehydration or acclimation to saline media, urea accumulates in their body fluids as a result of a combination of increased urea production and/or decreased glomerular filtration rates in the kidney (Goldstein, 1972; Funkhouser and Goldstein, 1973). The African clawed frog, *Xenopus laevis,* having been deprived of water and then placed back into water, excretes the urea stored in body fluids, and subsequently shifts to excreting ammonia (Balinski, 1970; Balinski et al., 1961). However, it is not clear how widespread this reverse shift is in other aquatic frogs. It may represent a highly specialized adaptation in *Xenopus.* Although *Xenopus* has been described by most investigators as an aquatic frog, when kept under conditions of low water availability, it exhibits many of the same physiological adaptations found in fossorial desert anurans. For example, *Xenopus* estivates underground for long periods of time when freshwater pools in which they live dry up (Rose, 1962). During this time they become ureotelic and urea accumulates in their body fluids (Balinski et al., 1967). Additionally, when kept under moderately saline conditions, *Xenopus* will increase the production and storage of urea (Funkhouser and Goldstein, 1973). Recently, *Xenopus laevis,* when acclimated to saline solutions between 570 and 625 mOsm/liter, was shown to increase its body fluid concentration to as much as 60 mOsm/liter greater than the medium primarily by storing urea in the body fluids (Schlisio et al., 1973). Therefore, *Xenopus* shows similar adaptive responses when estivating or when placed in hypersaline media. Since under both conditions the frog is faced with low water potentials and the possibility of water loss to the medium, the similarity of the responses is not surprising.

Since many aquatic frogs seem to be able to store urea in the absence of free water, it appears that anurans were preadapted to move into environments where the water potential was low, e.g., deserts and brackish water. Many fossorial desert anurans utilize this same basic pattern to survive underground for long periods. The only difference may be the extent to which arid-adapted anurans can produce and store urea. *Scaphiopus,* the spadefoot toad, is capable of storing up to 300 mM/l of urea in its body fluids while hibernating underground waiting for summer rains (McClanahan, 1967; Shoemaker et al., 1969). The rate at which urea is produced and stored is inversely proportional to soil water potential (McClanahan, 1972). If the soil in which they are burrowed dries and the soil water potential decreases, the toads store urea and lower their body water potential to prevent or slow cutaneous water loss to the soil. A recent study (Delson and Whitford, 1973) on the tiger salamander, *Ambystoma tigrinum,* an inhabitant of the Chihuahuan desert in southern New Mexico, showed that this species, after spending 9 months in underground burrows, stored urea and electrolytes in its body fluids at levels comparable to those of *Scaphiopus.* It is highly probable that *Ambystoma* also uses urea storage to offset the low water potentials that may occur in soils in which they live.

Not all arid-adapted fossorial anurans have exploited urea storage as a device to lower their body water potentials. An alternative would be to change the permeability of the skin to prevent water loss when soil water potential falls below that of body fluids, as in the South American ceratophryd frog, *Lepidobatrachus llanensis* (McClanahan et al., 1975). When kept in dry air or placed in dry soil, this normally aquatic frog will form a cocoon that is highly impermeable to evaporative water loss. Recent experiments in which *Lepidobatrachus* and *Scaphiopus* were placed in the same soil showed that *Scaphiopus* accumulated urea in the body fluids tenfold over normal hydrated concentrations. In contrast, *Lepidobatrachus* experienced only slight increases in plasma urea over aquatic levels. The impermeable cocoon in *Lepidobatrachus* obviates enhanced protein metabolism and urea storage.

Until recently it was believed that amphibians were incapable of excreting significant quantities of nitrogen in the form of uric acid or urate salts. Loveridge (1970) described the first arid-adapted African frog, *Chiromantis xerampelina,* that could excrete the majority of its nitrogen as urates. Shoemaker et al. (1972) discovered an Argentine hylid frog, *Phyllomedusa sauvagei,* that is also a uricotelic species. *Chiromantis xerampelina* is a member of the family Rhacophoridae, which is restricted to the Old World, and *P. sauvagei* belongs to the Hylidae, which is thought to have originated in the New World (Duellman, 1968; Schiotz, 1967; Poynton, 1964). Nevertheless, both species are arboreal and occupy similar ecological niches. Unlike other anurans that have been investigated, both have evaporative water losses comparable to those of desert reptiles (about 1 mg of H_2O g body weight $^{-1} \cdot h^{-1}$).

The advantages of being uricotelic and simultaneously having an integument that is relatively impermeable to water loss can be readily seen when an anuran such as *P. sauvagei* is compared with more mesic forms. We recently completed experiments in which total nitrogen excretion was assessed in several species of *Phyllomedusa* (*P. iherengi, P. pailona, P. hypochondrialis*); in a Central American hylid, *Agalychnis annae; Pachymedusa dacnicolor* from Mexico; and *Hyla pulchella,* a hylid species found associated with water in northern Argentina. *Phyllomedusa iherengi* and *P. pailona* occur in more tropical areas of northern Argentina; *P. hypochondrialis* occurs along the drainage of the Río Paraná. All these species afforded an excellent comparative study with *P. sauvagei* whose distribution extends into central Argentine deserts.

The frogs were placed in covered containers and kept at a constant temperature of 20°C. They were fed mealworms (larval *Tenebrio molitor*) at the rate of 1 percent of their body weight per day. No additional water was provided. *Phyllomedusa sauvagei* was maintained on this regime for 1 month with no apparent ill effects; *P. hypochondrialis* was also maintained 1 month, but showed signs of stress. The other species were maintained for various periods of less than 1 month. The bladders of experimental animals were drained before and after the experiment. Bladder urine collected at the termination of the experiment was analyzed for urate, urea, and ammonia. No animals voided urine during the experiment and all had bladder urine at the end of the experiment, except for *H. pulchella.* Plasma samples were taken at the beginning and end of the experiment and changes in electrolytes, urea, and ammonia were determined. After urine and plasma samples had been collected at the termination of the experiment, the frogs were placed in water for an additional 24 hours so that any urea, urate, or ammonia that had accumulated in the plasma could be excreted. The water was then analyzed for these excretory products.

The rate of nitrogen excretion in *P. sauvagei* was twice that of the other species of *Phyllomedusa* from Argentina, and urate excretion accounted for over 90 percent of the nitrogen excreted (Table 1). When given access to water for 24 hours after the feeding experiment, *P. sauvagei* excreted relatively little additional nitrogen over what had been excreted during the time without water. In the other species of *Phyllomedusa* considerably more urea and ammonia nitrogen accumulated in the body fluids during the experiment and were excreted when they were placed in water. In *A. annae* and *P. dacnicolor,* the nitrogen excretory rate was extremely low in the absence of water and most of the nitrogen was in the form of urates. However, significant amounts of urea were stored in their body fluids. When they were placed in water for 24 hours, urea excretion increased their total nitrogen excretory rates by as much as tenfold. The same, but more extreme, pattern is seen in *Hyla pulchella,* which excreted no nitrogen in the absence of water, but excreted large amounts of urea nitrogen when placed in water.

The efficiency of *P. sauvagei* at excreting nitrogen in the absence of water is better understood when water loss and rates of increase of plasma solutes are examined (Table 2). *Phyllomedusa sauvagei* lost the least amount of water of the anurans investigated and had the lowest rate of total solute increase. *Phyllomedusa pailona* and *P. iherengi* had higher rates of water loss and total solute increase than did *P. sauvagei.* Production and storage of urea in body fluids in these species contrib-

TABLE 1 Mean rates of nitrogen excretion in various anuran species during feeding trial with insect food as the only source of water

Species	Nitrogen excretion in absence of water (μg of Ng^{-1} day^{-1})				Nitrogen excretion in absence of water plus 1 day in water (μg of Ng^{-1} day^{-1})			
	Urate	Urea	NH₃	Total	Urate	Urea	NH₃	Total
Phyllomedusa sauvagei, N = 5	227	7.16	12.3	247	239	30.4	13.6	283
Phyllomedusa pailona, N = 5	106.3	11.3	12.3	130	100.4	75.1	13.4	189
Phyllomedusa iherengi, N = 3	110	18.4	5.8	134	104	69.6	13.6	186
Phyllomedusa hypochondrialis, N = 5	80	18.4	1.7	100	79.7	71.7	6.65	158
Agalychnis annae, N = 2	11	2.3	0.75	14.4	10.5	131	6.9	148.5
Pachymedusa dacnicolor, N = 4	7.6	5.5	2.5	15.6	7.2	94.1	5.1	106.5
Hyla pulchella, N = 3	0	0	0	0	0	491	31	523

TABLE 2 Mean rates of increase of plasma solutes and weight loss in various species of anurans during feeding trial with insect food as the only source of water

Species	Na⁺	K⁺	Cl⁻	Urea	Total	Weight loss
	(mmol/day)					(%/day)
Phyllomedusa sauvagei, N = 5	1.03	0.08	1.1	1.6	3.3	−0.03
Phyllomedusa pailona, N = 5	1.97	0.09	2.6	5.06	10.4	−0.092
Phyllomedusa iherengi, N = 3	1.36	0.31	2.8	5.8	11.9	−0.045
Agalychnis annae, N = 2	2.28	−0.17	5.4	13.4	23.5	−0.27
Pachymedusa dacnicolor, N = 4	3.30	−0.04	4.1	11.5	19.4	−0.32
Hyla pulchella, N = 3	7.5	0.35	7.8	26.4	40.0	−2.7

uted significantly to the total solute change during the experiment. Rates of water loss in *A. annae* and *P. dacnicolor* were 10 times those of *P. sauvagei*. This is reflected in the rate of increase of individual and total solutes. *Hyla pulchella* showed an approximate 100-fold increase in rate of water loss over *P. sauvagei* and elevated rates of change in most plasma solutes were observed.

The inability of the more aquatic anurans investigated to excrete significant amounts of nitrogen in the absence of water is probably related to high rates of evaporative water loss and consequent cessation of kidney function. For example, the glomerular filtration rate and urine flow in frogs drop precipitously as they are dehydrated in air (Schmidt-Nielsen and Forster, 1954; Schmidt-Nielsen, 1972).

We have examined the activities of some key enzymes in the ornithine and purine cycles for several uricotelic and nonuricotelic anurans and for the desert lizard, *Dipsosaurus dorsalis* (Table 3) (Shoemaker and McClanahan, unpublished data). Although *P. hypochondrialis* and *P. sauvagei* have more arginase activity in their livers than *D. dorsalis*, they nevertheless have much lower liver arginase activity than the aquatic frog *Leptodactylus chaquensis* and the more mesic toad *Bufo arenarum*. Furthermore, *L. chaquensis* and *B. arenarum* both have low xanthine oxidase activity in the kidney and liver whereas *P. hypochondrialis* and *P. sauvagei* have liver and kidney xanthine oxidase activities comparable to those in the kidney of *D. dorsalis*. Uricase activities were found to be low in the livers and kidneys of the uricotelic frogs and the desert lizard and high in the ureotelic frog and toad. These data demonstrate that, although the distribution of purine cycle enzymes in the livers and kidneys of uricotelic frogs and lizards is different, the activity levels are comparable.

In contrast to adult amphibians, larval amphibians have been considered to be ammonotelic (Balinski, 1970), shifting to ureotelism only at metamorphosis. Many studies have shown that the shift to ureotelism at metamorphosis is accompanied by increased activities of urea cycle enzymes (Cohen and Brown, 1960; Brown et al., 1959, reviewed by Balinski, 1970). However, the generalizations are based on observations of a few species that have primarily aquatic larvae. Very few studies have been completed on larval forms living under conditions of restricted water. The anuran family Leptodactylidae represents an interesting group of frogs because some of its members deposit their eggs to hatch in areas devoid of free water. The larvae subsequently enter temporal pools of water to develop. There is strong evidence that larval ureotelism may be widespread in this family. Candelas and Gomez (1963) reported storage of urea, but not ammonia, in tissues of larval *Leptodactylus albilabrus* when they were exposed to "nonaquatic" conditions, whereas ammonia accumulated in the tissues of *Rana*

TABLE 3 Activities[a] of arginase, xanthine oxidase, and uricase in liver and kidney of *Phyllomedusa sauvagei* compared with other species of frogs and a lizard *(Dipsosaurus dorsalis)*

Species	N	Arginase ($\mu mol\ mg^{-1} \cdot h^{-1}$)		Xanthine oxidase ($m\mu mol\ mg^{-1} \cdot h^{-1}$)		Uricase ($m\mu mol\ mg^{-1} \cdot h^{-1}$)	
		liver	kidney	liver	kidney	liver	kidney
Phyllomedusa	6	16.9	0.32	62.0	69.3	11.0	9.7
sauvagei		(7.2–28)	(BDL[b]–0.6)	(41–77)	(45.5–94.0)	(7.0–22)	(BDL–17)
Phyllomedusa	4	14.8	1.4	22.4	79.5	7.4	1.1
hypochondrialis		(12–17)	(1.3–1.4)	(15–41)	(74–85)	(2.7–13)	(BDL–2.2)
Leptodactylus	5	89.1	1.1		23.1	70.2	61.1
chaquensis		(71–105)	(0.7–1.6)	BDL	(17–29)	(42–97)	(39–95)
Bufo arenarum	4	38.8	1.2		38.3	87.6	30.0
		(20–73)	(0.2–3.3)	BDL	(33–48)	(78–117)	(19–44)
Dipsosaurus	6	<1	1.8		56.7	25.5	18.8
dorsalis			(1–3)	BDL	(30–75)	(20–33)	(17–20)

[a] Rate of product formation or substrate disappearance per unit of tissue fresh weight.
[b] Below detectable levels using methods described.

catesbiana tadpoles kept under the same conditions. Another Australian leptodactylid, *Crinia victoriana,* has larvae that may remain for long periods on land enclosed in egg capsules. These larvae also contain considerably more urea than ammonia in their body fluids (Martin and Cooper, 1972).

Frogs of the Cavicola group of the genus *Leptodactylus* deposit their eggs in mud burrows excavated near the margins of seasonal ponds (Cei, 1956). The only moisture available is that in the proteinaceous foam that is deposited along with the eggs and whipped up by the amplexing pair (Cei, 1949). The larvae may live in the foam for several weeks awaiting rain, at which time the pond level reaches the nest and the larvae resume development in the water. Shoemaker and McClanahan (1973) found that larvae of *Leptodactylus bufonius,* a species found in the Argentine chaco, are ureotelic while in the nest and during development in water (Fig. 1). Urea accounts for approximately 90 percent of the total nitrogen excreted by the tadpoles in the foam nest. After transferring the tadpoles to water, the fraction of urea decreased to 77 percent and remained at that level as long as the tadpoles were not fed. After feeding, total nitrogen output decreased and urea and ammonia accounted for equal proportions of the nitrogen excreted until metamorphosis. At that time, total nitrogen excretion increased dramatically and urea nitrogen accounted for 80 to 85 percent of the total. Thus, fluctuations in the total nitrogen excreted during development were mirrored mainly by fluctuations in urea excreted. The lethal limit of urea tolerance for tadpoles is between 450 and 500 mM, and ammonium

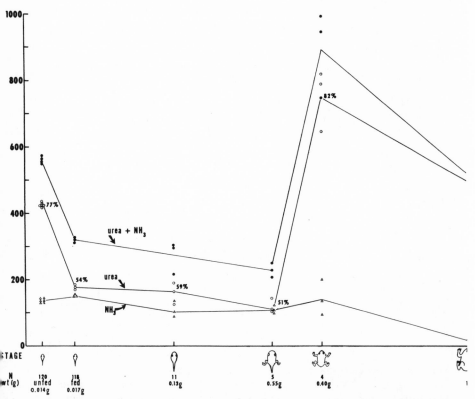

FIGURE 1 Rates of excretion of urea (o), ammonia (△), and urea plus ammonia (●) ...otodactylus bufonius at various stages. Urea N as a percentage of total is indicated al... ...e showing urea excreted. Stage, number of individuals used, and their average wei... ...e indicated on the abscissa. Spacings on abscissa call attention to time differences ...e not strictly proportional to them. (From Shoemaker and McClanahan, 1973.)

chloride concentrations as low as 5 mM resulted in 100 percent mortality of experimental animals.

The adaptive significance of ureotelism in the larval stages in forms such as *L. bufonius* is readily apparent. It would be of interest to examine other leptodactylids whose larvae are fully aquatic and other amphibians whose eggs and larvae may be forced to develop in situations where there is little free water.

The ontogeny of uricotelism has been investigated in larval *P. sauvagei* (Shoemaker and McClanahan, unpublished data). The tadpoles were not principally ammonotelic at any of the developmental stages examined (Fig. 2). There was essentially no urate nitrogen produced until forelimb emergence. At all developmental stages prior to metamorphosis, urea nitrogen excreted was always slightly greater than

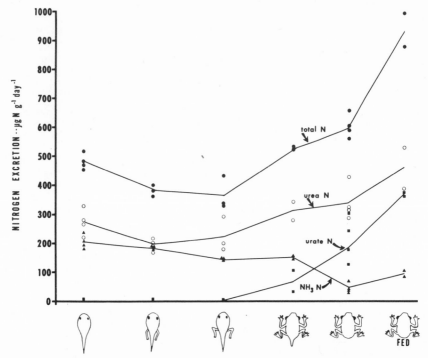

FIGURE 2   Rates of excretion of urea (○), ammonia (▲), urate (■), and total N (●) in *Phyllomedusa sauvagei* at various stages. Spacings on abscissa call attention to time differences but are not strictly proportional to them.

ammonia nitrogen excreted. Ammonia nitrogen did not drop appreciably until metamorphosis. At this time both urea nitrogen and urate nitrogen excreted increased, with urea nitrogen comprising 58 percent and urate nitrogen comprising 25 percent of the total. In newly metamorphosed frogs that had fed, the urea nitrogen represented 47 percent of the total nitrogen excreted, and urate nitrogen represented 37 percent. Presumably, they do not become principally uricotelic until after they have left the water and become arboreal. In light of these findings if would be instructive to look at other arboreal hylids to find out if they are ureotelic during the aquatic larval stages.

From data presented here it is obvious that patterns of nitrogen excretion within the amphibia are as diverse as those found in many reptiles. Consequently, we must discount the idea that in the evolution of vertebrates there was a phylogenetic transition from ammonotelism to uricotelism (Baldwin, 1959). It is more plausible that within the amphibia those forms that survived in arid environments became proficient at excreting nitrogen as urea or urate. Thus far, uricotelism has been described only in arboreal anurans that must remain exposed to high temperatures and potentially low humidities. It has been found in several tropical species as well as arid-adapted species (Table 1). This suggests that uricotelism may have evolved in tropical forms but became particularly effective for water conservation in those arid-adapted forms which also evolved a more effective mechanism to prevent evaporative water loss. In fact, feeding experiments with various hylid species (Table 1) in the absence of water indicate that there is no particular advantage in being uricotelic unless evaporative water loss is quite low. The latter is atypical of most amphibians. In contrast, ureotelism has been exploited in many desert forms that behaviorally avoid the extremes of the environment by burrowing into the soil and remaining inactive for long periods of time.

A word of caution seems in order to environmental physiologists on the basis of the information in this review. We often tend to focus our interests on the most readily available species, and in many cases generalize our conclusions for a whole group. This procedure disregards the diversity within the group and the likelihood of a great number of patterns of adaptability, and tends to discourage others from doing research on potentially productive problems that could shed information on this diversity.

Acknowledgments

The research of McClanahan and Shoemaker cited in this review was supported by NSF Grants GB-19084 and GB-29604.

LITERATURE CITED

Baldwin, E. 1959. Dynamic aspects of biochemistry. Cambridge University Press, New York.

Balinski, J. B. 1970. Nitrogen metabolism in amphibians. In J. W. Campbell (ed.), Comparative biochemistry of nitrogen metabolism, pp. 519–624. Academic Press, New York.

———, E. L. Choritz, C. G. L. Coe, and S. Van Der Schans. 1967. Amino acid metabolism and urea synthesis in naturally aestivating *Xenopus laevis*. Comp. Biochem. Physiol. 22:59–68.

———, M. Cragg, and E. Baldwin. 1961. The adaptation of amphibian waste nitrogen excretion to dehydration. Comp. Biochem. Physiol. 3:236–244.

Brown, G. W., Jr., W. R. Brown, and P. P. Cohen. 1959. Levels of urea cycle enzymes in metamorphosing *Rana catesbiana* tadpoles. J. Biol. Chem. 234:1775–1780.

Campbell, J. W., and L. Goldstein. 1972. Nitrogen metabolism and the environment. Academic Press, New York.

Candelas, C., and M. Gomez. 1963. Nitrogen excretion in tadpoles of *Leptodactylus albilabrus* and *Rana catesbiana*. Amer. Zool. 3:521–522 (abstr.).

Cei, J. M. 1949. Costumbres nupciales y reproducción de un batracio característico chaqueno (*Leptodactylus bufonius* Boul.). Acta. Zool. Lilloana (Tucuman, Argentina) 8:105–110.

————. 1956. Nueva lista sistematica de los batracios de Argentina y breves notas sobre su biología y ecología. Invest. Zool. Chilenas 3:35–68.

Cohen, P. P., and G. W. Brown. 1960. Ammonia metabolism and urea biosynthesis. In M. Florkin and H. S. Mason (eds.), Comparative biochemistry, Vol. 2, pp. 161–244. Academic Press, New York.

Delson, J., and W. G. Whitford. 1973. Adaptations of the tiger salamander, *Ambystoma tigrinum*, to arid habitats. Comp. Biochem. Physiol. 46A:631–638.

Duellman, W. E. 1968. The genera of phyllomedusine frogs. Publ. Mus. Nat. Hist. Univ. Kans. 18:1–10.

Funkhouser, D., and L. Goldstein. 1973. Urea response to pure osmotic stress in the aquatic toad, *Xenopus laevis*. Amer. J. Physiol. 224(3):524–529.

Goldstein, L. 1972. Adaptation of urea metabolism in aquatic vertebrates. In J. W. Campbell and L. Goldstein (eds.), Nitrogen metabolism and the environment, pp. 55–77. Academic Press, New York.

Loveridge, J. P. 1970. Observations on nitrogenous excretion and water relations of *Chiromantis xerampelina* (Amphibia, Anura). Arnoldia 5(1):1–6.

Martin, A. A., and A. K. Cooper. 1972. The ecology of terrestrial anuran eggs, genus *Crinia* (Leptodactylidae). Copeia 1972:163–168.

McClanahan, L., Jr. 1967. Adaptations of the spadefoot toad, *Scaphiopus couchii* to desert environments. Comp. Biochem. Physiol. 20:73–99.

————. 1972. Changes in body fluids of burrowed spadefoot toads as a function of soil water potential. Copeia 1972:209–216.

————, V. H. Shoemaker, and R. Ruibal. 1974. Structure and function of the cocoon of a ceratophyrd frog. Copeia (in press).

Poynton, J. C. 1964. The amphibia of Southern Africa. Ann. Natal. Mus., Vol. 17. Pietermaritzburg.

Rose, W. 1962. The reptiles and amphibians of Southern Africa, 2nd ed. The Standard Press Limited, Cape Town, South Africa.

Schiotz, A. 1967. The tree frogs (Rhacophoridae) of West Africa. Spolia. Zool. Mus. Haunensis 25:1.

Schlisio, V. W., K. Jurss, and L. Spannhof. 1973. Osmo- und Ionenregulation von *Xenopus laevis* DAUD. nach adaptation in verschiedenen osmotisch wirksamen Lösungen. Zool. Jb. Physiol. Bd. 77:275–290.

Schmidt-Nielsen, B. 1972. Urea excretion by the vertebrate kidney. In J. W. Campbell and L. Goldstein (eds.), Nitrogen metabolism and the environment, pp. 81–103. Academic Press, New York.

————, and R. P. Forster. 1954. The effect of dehydration and low temperature on renal function in the bullfrog. J. Cell. Comp. Physiol. 44:233–246.

Shoemaker, V. H., D. Balding, R. Ruibal, and L. McClanahan. 1972. Uricotelism and low evaporative water loss in a South American frog. Science 175:1018–1020.

————, and L. McClanahan. 1973. Nitrogen excretion in the larvae of a land-nesting frog (*Leptodactylus bufonius*). Comp. Biochem. Physiol. 44:1149–1156.

————, L. McClanahan, Jr., and R. Ruibal. 1969. Seasonal changes in body fluids in a field population of spadefoot toads. Copeia 1969:585–591.

THE WATER RELATIONS OF TWO POPULATIONS OF NONCAPTIVE DESERT RODENTS

Richard E. MacMillen and Ernest A. Christopher

Department of Population and Environmental Biology, University of California, Irvine, California, and Department of Biology, Santa Rosa Junior College, Santa Rosa, California

Abstract

A field study of urine and blood osmotic properties of nocturnal desert rodents during a 3-year period revealed repeating annual patterns that appeared to be independent of precipitation during drought and nondrought years. For the granivorous heteromyids *Dipodomys merriami, Perognathus longimembris,* and *P. fallax,* urine concentrations were directly related to mean minimal air temperatures, which is predictable from the relationship between metabolic water production–evaporative water loss and ambient temperature for water-independent rodents. For the water-dependent cricetids *Neotoma lepida* and *Onychomys torridus,* urine concentrations were uniformly low and high, respectively, reflecting the solute loads of their diets (herbivorous and carnivorous, respectively). The cricetid *Peromyscus crinitus,* which is approaching water independence, appears equally suited to subsistence on a carnivorous or a granivorous diet, and its urine concentrations suggest dietary switching. All rodents studied appeared to be in positive states of water balance even during a period of severe drought, suggesting that they are seldom faced directly with the problem of water stress.

INTRODUCTION

Voluminous data have been gathered in the laboratory on the water metabolism of desert rodents, with extrapolative and interpolative interpretations based on the probable environmental stresses imposed upon the experimental animals by their natural habitats. Typical of such data are the abundant demonstrations of maximal urine concentrating capacities of desert rodents under conditions of osmotic stress, which appear to be indicative of relative degrees of renal water conservation. Less attention has been paid to laboratory assessments of fecal and evaporative water loss, even though the latter represents approximately 70 percent of the total water loss in the kangaroo rat, *Dipodomys merriami* (Schmidt-Nielsen, 1972). MacMillen (1972) provides some recent comparative data on renal concentrating capacity and on fecal and evaporative water loss in desert rodents.

Almost no attention has been paid to assessments of states of water balance of desert rodents under field conditions, and the frequency with which water conserving mechanisms are actually used. In this paper we examine critically the relationships between the two major avenues of water loss—evaporative and renal—and apply these relationships to a series of field measurements of blood and urine osmotic properties of two populations of desert rodents. Further, this examination includes a year of normal rainfall, followed by a year of drought, and then a subsequent year of renewed water availability. A brief, preliminary report of this study is included in MacMillen (1972).

MATERIALS AND METHODS

The Habitats

Both sites employed in this study are located on the lower slopes of the Little San Bernardino Mountains, San Bernardino County, California, in desert habitats that are transitional between the Sonoran and Mojave Deserts.

From November 1970 to December 1972 the study site was in a small desert valley bounded by granitic ridges located about 3.3 km south of Joshua Tree, California, on the north (inland) aspect of the Little San Bernardino Mountains. Elevation at the site is approximately 970 m, and mean annual rainfall, recorded 26 km to the east at Twenty-Nine Palms, approximates 103 mm. Annual rainfall for the 3 years of the study is indicated in Table 1.

In February 1973, because of severe drought and near disappearance of the rodent fauna at Joshua Tree, the study site was relocated to Morongo Valley, situated at the northwestern extremity of the Little San Bernardino Mountains at an elevation of 830 m. This valley is exposed somewhat to coastal climatic influences, and hence has a greater rainfall and was less affected by the drought. Annual rainfall for 1972 and 1973 in Morongo Valley exceeded that of Joshua Tree by five and four times, respectively (Table 1).

TABLE 1 Annual rainfalls in Joshua Tree and Morongo Valley, California, 1971–1973

Locality	Rainfall (mm)		
	1971	1972	1973
Joshua Tree[a]	103	21	39
Morongo Valley[b]	—	107	156

[a] From Monument Headquarters, Joshua Tree National Monument, Twenty-Nine Palms, Calif.
[b] From data collected by Jack Frances at Morongo Valley, Calif.

Both sites comprise very similar habitats made up of a mixture of creosote bush scrub and Joshua tree woodland. The dominant perennial vegetation was creosote bush (*Larrea divaricata*), Joshua tree (*Yucca brevifolia*), blackbush (*Coleogyne ramosissima*), boundary ephedra (*Ephedra aspera*), and perennial bunchgrass (*Hilaria rigida*). During spring, desert annuals were variably present, but were particularly abundant during spring 1973.

The Rodents

This report deals only with the nocturnally active rodents at the two sites. These include the heteromyid rodents *Dipodomys merriami, Perognathus fallax,* and *P. longimembris,* and the cricetids *Neotoma lepida, Onychomys torridus,* and *Peromyscus crinitus.* Representative mean body weights of these species are presented in Table 2. No *N. lepida* or *P. crinitus* were taken at the Morongo Valley site.

Trapping Techniques

From November 1970 to December 1972 trapping was conducted once each month during the dark of the moon at the Joshua Tree site. A single, medium-sized Sherman live trap was set at each of 100 permanent trapping stations arranged either in lines or in a quadrant; traps were baited with mixed birdseed. Traps were arranged to sample both the rodents occurring on the desert floor (principally *D. merriami, O. torridus,* and *P. longimembris*) and in the bordering granitic outcrops (*N. lepida, P. crinitus,* and *P. fallax*). Traps were set just before dusk, and checked at midnight and again at 0600. The first 12 individuals of each species captured were transported to a nearby cottage for collection of blood and urine. All animals were marked by toe clipping for identification, and were then released at their sites of capture by 1800 on the day

TABLE 2   Body weights of the rodent species

	D. merriami	P. fallax	P. longi- membris	N. lepida	O. torridus	P. crinitus
Mean body weight (g)	37.3	21.3	9.0	125.5	20.2	14.5
Standard deviation	4.1	2.5	0.4	30.8	2.1	1.1
Number of individuals	10	6	7	9	6	9

following the night of capture. Prior to release the animals were pro
vided with mixed birdseed for food, but no access to water or othe
succulent foods.

In 1973 the study site was relocated to Morongo Valley and trappinç
was conducted on 100 permanent trapping stations arranged in a quad
rat, twice a month during alternate months commencing February 1973
The twice-a-month schedule was necessary for injecting tritium anc
reanalyzing its activity prior to complete exchange. Traps were se
before dusk and baited with mixed birdseed. Traps were checked anc
then closed at midnight, and all animals captured were returned to the
nearby cottage for urine and blood collection, and prepared for tritiun
analysis.

Collection and Analysis of Blood and Urine

Within 1 hour after removal from traps and within no more than
hours after capture the animals were placed in urine collecting cages
These were constructed from 1-lb coffee cans with hardware-clotl
bottoms and suspended over a dish filled with mineral oil which caugh
urine as it was voided. For protection against predators during urine
collection, the animals were housed in a small metal shed; the shed wa:
neither heated nor insulated and its interior approximated surface con
ditions of temperature. The animals were not provided with food durinç
the 4-hour period of urine collection. The urine was transferred b
pipette to small vials, under a cover of mineral oil, and frozen unti
analysis.

Animals were etherized after urine collection, and blood sample:
were taken, using heparinized capillary tubes, from the infraorbita
sinus. Typically, two or three 80-μl samples were removed from eacl
animal. The blood was centrifuged, and the plasma frozen in sealec
capillary tubes for later analysis.

This report concentrates on measurements of blood and urine osmo
tic pressures, which were determined with a Mechrolab vapor pressure
osmometer in the laboratory at the University of California at Irvine. Ir
addition, measurements were routinely made of blood and urine ure&
concentrations using the Conway microdiffusion technique, chloride

ɔncentrations using an Aminco-Cotlove chloride titrator, and blood ematocrits.

So that field data could be correlated with more conventionally de-ʋed laboratory data for the desert rodents, urine measurements were so made in the laboratory with animals fully hydrated and water-ɘprived. Samples of each species were maintained individually in plas-ɔ cages with sand-covered bottoms and hardware-cloth lids, and ɔused in a windowless animal room on a 12-hour photoperiod. The ʌimals were provided with mixed birdseed for food, and were main-ined at an ambient temperature of 21 to 25°C and 36 to 57 percent r.h. ydrated animals were provided with fresh apple as a water source; ater-deprived animals had only air-dried seeds (water content about) percent by weight). Urine was collected overnight using the same ιethod as employed in the field.

ater Turnover

Turnover of body water was estimated using tritiated water. Animals ɘre injected intraperitoneally with 10µl of tritiated water (specific ɔtivity 10 mCi/ml). Following the method of Holleman and Dieterich 973), 3 hours were allowed for equilibration after injection, after which ʌ initial blood sample was removed from the infraorbital sinus. After ɘntrifugation, the blood sample was sealed and frozen in a heparinized apillary tube for later analysis. A second sample was removed and ɘated in the same manner from each animal that was recaptured ɘfore 1 month had elapsed since the initial treatment with tritium; sually the time elapsed was 1 week.

Tritium activity of the blood samples was determined at the University ɟ California at Irvine with a Beckman CPM 100 or LS 100 Liquid cintillation Counter, after the method of Minnich and Shoemaker 970). Ten microliters of plasma were added to scintillation vials con-ιining 10 ml of scintillation fluid. The fluid was prepared by adding 100 of naphthalene and 5 g of 2,5-diphenyloxazole, and then filling to 1 ʈer with "Baker analyzed" 1,4-dioxane. Count accuracy was to ±1 ɘrcent error, and counts were corrected for background and quench-g.

The rate constant (k), the fraction of body water exchanged per day, as calculated from the logarithmic decay function:

$$\text{final activity} - \text{initial activity} = e^{-kT}$$

here $\qquad k = \dfrac{\text{initial activity} - \text{final activity}}{\text{time in days}}$

ʌis is after the method of Holleman (1972), and expresses turnover as ɩe fraction of body water exchanged per day.

RESULTS

Climatic and Population Fluctuations

Seasonal distributions of mean monthly air temperatures and of monthly precipitation for the duration of this study at the two sites are indicated in Figure 1. Mean monthly maxima and minima air temperatures showed a repeating annual pattern, with midsummer highs and midwinter lows occurring during the same months each year. Of particular importance to this study (see Discussion) was the observation that mean monthly temperature minima exceeding 16°C were confined to the midsummer months of June–September; of the available air temperature data, we feel that the mean monthly minima best approximate the nocturnal conditions of surface temperatures encountered by the rodents while active on the surface at night.

Precipitation during the 3 years was very erratic, both in absolute amounts (Table 1) and in seasonal distribution (Fig. 1). In 1971 at Joshua Tree, an average rainfall year with 103 mm, there was very little winter rain and considerable midsummer rain. This pattern resulted in a rather poor germination and subsequent bloom of desert annuals. In 1972, a year of extreme drought with virtually no annual plant germination, rainfall was confined to October and November. In 1973, even though annual precipitation was below normal in Joshua Tree, most of the rainfall at both Joshua Tree and Morongo Valley occurred in January, February, and March, and was responsible for a massive germination and bloom of desert annuals that spring at both sites.

The fluctuations of nocturnal rodents at the two sites was also erratic (Fig. 1). During 1971 there was a pronounced population cycle, with minimum abundance in midwinter and maximum abundance in midsummer. In 1972 the population barely rose above the December 1971 low level, remained fairly stable at this reduced level throughout the summer, and then declined sharply in November and December. Commencing in February 1973, trapping success at Morongo Valley was considerably greater than it had been 2 months previously in Joshua Tree, and the population continued to increase throughout the spring and summer, finally showing a decline in November. Casual sampling at the Joshua Tree site indicated a recovery of that population during 1973, also.

Field Urine and Blood Osmotic Properties

Urine and blood concentrations, expressed as osmotic pressure, of *Dipodomys merriami* and *Peromyscus crinitus* collected between November 1970 and November 1973 are indicated in Figure 2. As no *P.*

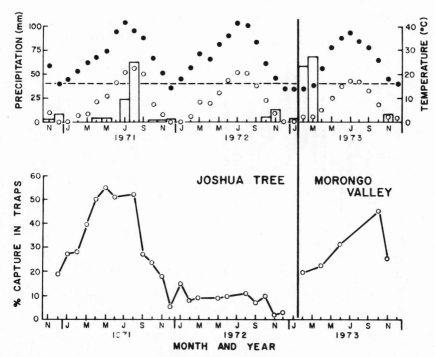

FIGURE 1 Monthly precipitation, mean monthly maximum and minimum temperatures (upper figure), and relative population fluctuations of nocturnal rodents (lower figure) at the two study sites, Joshua Tree and Morongo Valley. In the upper figure, vertical bars represent monthly precipitation, solid circles represent mean monthly temperature maxima, and open circles represent mean monthly temperature minima; the horizontal dashed line indicates a temperature of 16°C. In the lower figure the % capture in traps represents the proportion of 100 traps containing rodents at the first trap check (midnight) of each monthly trapping period.

crinitus were trapped in Morongo Valley during 1973, all data for that species are derived from the Joshua Tree site. In *D. merriami* there was a pronounced annual cycling of urine concentrations, which followed closely and directly the seasonal cycles of temperature (see Fig. 1): lowest concentrations invariably occurred in midwinter, averaging around 1000 mOsm/liter, while highest concentrations (significantly higher than midwinter values) occurred during midsummer and averaged around 4000 mOsm/liter. This cycle was repeated over all 3 years. *Peromyscus crinitus* had urine concentrations that routinely were as high (midsummer) or higher (midwinter) than those of *D. merriami,* and with much less pronounced annual cycling.

Osmotic pressure of blood plasma of *D. merriami* and *P. crinitus* was stable and comparable throughout the duration of the study, generally

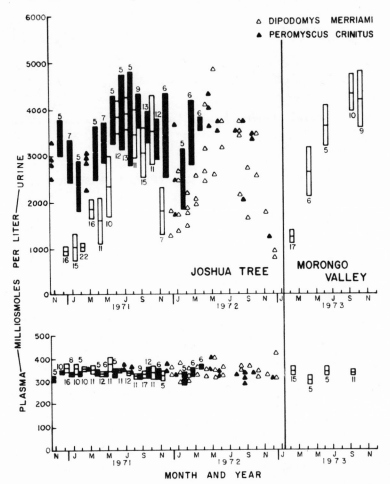

FIGURE 2   Osmotic pressure concentrations of urine (upper figure) and plasma (lower figure) of *Dipodomys merriami* and *Peromyscus crinitus*. Vertical bars represent the statistic $\bar{X} \pm t_{.95}$ S.E. (95 percent confidence limits), for sample sizes of five or larger; the horizontal lines in the bars represent means. Triangles represent individual measurements. Numbers near the bars represent sample sizes.

varying around a mean of 350 mOsm/liter for both species. Although variations did occur, these were in no way cyclic and therefore did not correspond to the annual cycles of urine osmotic pressure.

Urine and blood osmotic concentrations of *Perognathus longimembris* and *Onychomys torridus* are shown in Figure 3. No data are available for *P. longimembris* for November–February, as this species exhibits winter dormancy and is not active (trappable) on the surface. Although the data are seasonally truncated because of winter dormancy, *P. lon-*

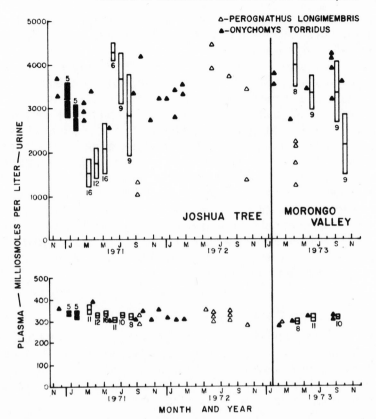

FIGURE 3 Osmotic pressure concentrations of urine (upper figure) and plasma (lower figure) of *Perognathus longimembris* and *Onychomys torridus*. For the two sets of urine concentrations of *P. longimembris* for April 1973, the four triangles represent data from the first week and the bar represents data from the second week. Symbols as in Figure 2.

gimembris also has urine concentrations that cycle on an annual basis, with lowest concentrations (about 1200 to 2000 mOsm/liter) just after emergence from dormancy in March–April and just prior to disappearance in September–October. With a single exception, highest urine osmotic concentrations (averaging > 3500 mOsm/liter) in *P. longimembris* occurred in midsummer. This exception occurred in April 1973 when values rose from a mean of 1800 mOsm/liter in the first week to a significantly ($p < 0.05$) higher mean of 4000 mOsm/liter in the second week. This sudden increase in urine concentration occurred during a period of lush annual vegetative production and was coincidental with a sudden appearance of large numbers of painted lady butterfly (*Pyrameis cardui*) larvae.

Although data are sparse (as was the mouse), *Onychomys torridus* had urine osmotic concentrations that were fairly high and stable throughout the study period in both habitats, generally ranging between 3000 and 4000 mOsm/liter (Fig. 3).

Blood plasma osmotic concentrations in both *P. longimembris* and *O. torridus* were also stable and comparable throughout the study period, ranging generally around 350 mOsm/liter (Fig. 3).

Urine and blood osmotic pressures are depicted in Figure 4 for *Perognathus fallax* and *Neotoma lepida* in the field. *Perognathus fallax* showed annual cycling in urine concentrations, again corresponding to the annual temperature cycles, with midwinter concentrations of around 1200 mOsm/liter and midsummer concentrations generally exceeding 4000 mOsm/liter. *Neotoma lepida* was not trapped in Morongo Valley; in Joshua Tree its urine concentrations remained quite dilute and fairly constant, generally ranging between 1000 and 2000 mOsm/liter. Blood plasma osmotic concentrations of *P. fallax* and *N. lepida,* as for the other species, were fairly stable and comparable throughout the study, ranging around 350 mOsm/l (Fig. 4).

Laboratory Urine Osmotic Properties

Mean urine concentrations of each of the species maintained in the laboratory and provided with a diet of mixed birdseed for food and apple for water (hydrated) or with only the birdseed (water-deprived) are indicated in Table 3. Since the heteromyids would not use the apple, only water-deprived data are presented for them.

Water Turnover

Data on exchange of body water for *D. merriami* and *P. fallax* at the Morongo Valley site are shown in Figure 5 and demonstrate an annual periodicity in rates of water turnover, with highest rates of water exchange during the cooler winter months and lowest rates of exchange during the warm, dry summer months.

DISCUSSION

Expressed as a function of oxygen consumption, the rates of evaporative water loss in rodents are directly related to ambient temperature. MacMillen and Grubbs (in press) have recently demonstrated that, for those rodents for which comparable data are available, rates of evaporative water loss of desert rodents (8 species, 5 genera, 3 families) are not statistically different from those of nondesert rodents (14 species, 9 genera, 4 families) at the same ambient temperature (Fig. 6). MacMillen and Grubbs (in press) have transformed the data from Figure

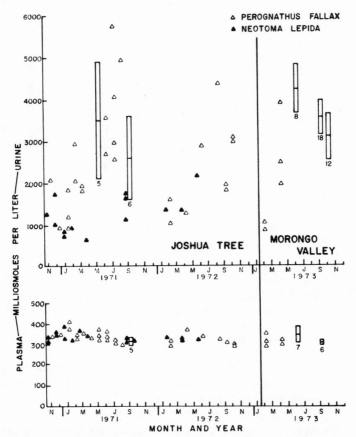

FIGURE 4   Osmotic pressure concentrations of urine (upper figure) and plasma (lower figure) of *Perognathus fallax* and *Neotoma lepida*. Symbols as in Figure 2.

TABLE 3   Urine osmotic concentrations of hydrated and water-deprived desert rodents under laboratory conditions[a]

	Hydrated (mOsm/liter)			Water-deprived (mOsm/liter)		
	$\overline{X}$	±S.D.	N	$\overline{X}$	±S.D.	N
Dipodomys merriami	—	—	—	3165	417	10
Neotoma lepida	396	236	10	2436	445	9
Onychomys torridus	1890	590	6	2733	262	5
Perognathus fallax	—	—	—	1241	420	8
Perognathus longimembris	—	—	—	1675	403	8
Peromyscus crinitus	2524	575	11	3047	436	8

[a] Ambient temperature fluctuated between 21 and 25°C, and relative humidity between 36 and 57 percent.

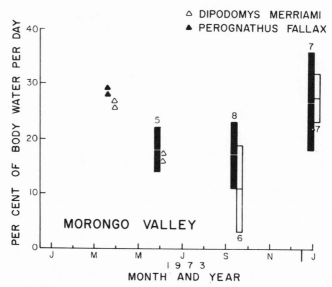

FIGURE 5 Daily rate of exchange of total body water in *Dipodomys merriami* and *Perognathus fallax* at Morongo Valley, as determined by the rate of loss of injected tritium. Symbols as in Figure 2.

6 into the relationship between metabolic water production–evaporative water loss and ambient temperature by assuming that all the rodents for which data were available were oxidizing pearled barley as an energy source, with a yield of 0.64 mg of metabolic water for each 1.0 cm³ of oxygen consumed (Fig. 7). Figure 7 shows an indirect relationship between the ratio of metabolic water production (MWP) to evaporative water loss (EWL) and ambient temperature (t_a) for rodents in general, and predicts that MWP = EWL at t_a = 16.6°C. For rodents operating at t_a < 16.6°C, MWP > EWL, and for rodents operating at t_a > 16.6°C, MWP < EWL. Hence, for nocturnally active seed-eating rodents operating under cool to cold conditions, as exemplified by the granivorous rodents studied during the winter months (see Fig. 1), the animals could well be faced with a surplus of metabolic water production, resulting in minimal urine concentrations. Conversely, nocturnal surface activity during the warmer summer months would result in a less favorable relationship between MWP and EWL, requiring greater water conserving capacities including maximal urine concentrations.

It should be emphasized, however, that Figure 7 indicates considerable interspecific variability, even though this form of comparison reveals no differences between desert and nondesert forms. Thus, for certain rodents the ratio of MWP to EWL is much greater at a given temperature than for others from the same general habitat. The only

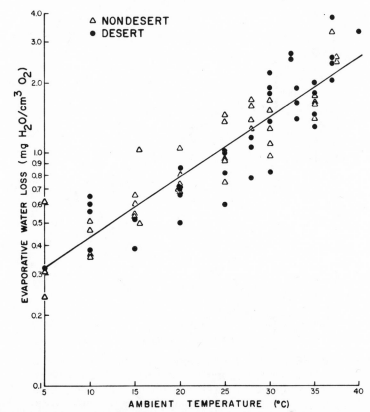

FIGURE 6 Relationship between evaporative water loss (EWL) and ambient temperature (t_a) in 8 species of desert rodents and 14 species of nondesert rodents. Each filled circle represents a mean measurement for a sample of individuals of a desert species; each open triangle represents a mean measurement for a sample of individuals of a nondesert species. The diagonal line is fitted to the pooled data by the least squares equation: log EWL = $-0.624 + 0.026t_a$. (After MacMillen and Grubbs, in press.)

species for which data are available (Raab and Schmidt-Nielsen, 1972; Carpenter, 1966) and which was studied by us is *Dipodomys merriami,* a granivorous heteromyid. *Dipodomys merriami* has the highest ratio of MWP to EWL of any of the included rodents at t_a = 15, 20, 25, and 30°C (Fig. 7), and a line fitted by eye to these data indicates that in *D. merriami* MWP = EWL at $t_a \approx 25°C$.

The interspecific variability in MWP–EWL and t_a for desert rodents as indicated in Figure 7 suggests the probability of equal variability within a particular population of sympatric desert rodents. This is precisely the situation observed when the rodent species we studied are maintained without water on an air-dry diet in the laboratory at t_a = 21 to 25°C. The

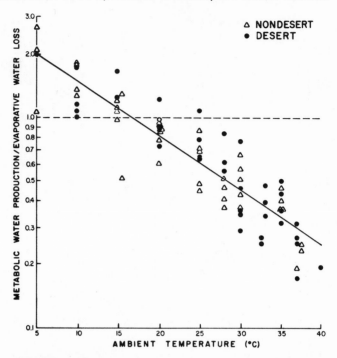

FIGURE 7 Relationship between metabolic water production (MWP)/evaporative water loss (EWL) and ambient temperature (t_a) in desert and nondesert rodents. Symbols as in Figure 6. The diagonal line is fitted to the pooled data by the least squares equation: log MWP–EWL = 0.430 − 0.026t_a. (After MacMillen and Grubbs, in press.)

six species form a gradient of weight response to water deprivation, with the typically granivorous heteromyids maintaining or gaining weight and the typically omnivorous to carnivorous cricetids losing weight to varying degrees (Fig. 8). Further, the relationship between MWP and EWL at these laboratory temperatures (Fig. 7) is near the break-even point for *D. merriami;* weight maintenance (Fig. 8) plus relatively high urine concentrations (Table 3) while water-deprived verify this contention. The ability of *Perognathus fallax* and *P. longimembris* to gain weight on a dry-seed diet (Fig. 8) and still produce very dilute urine (Table 3) suggests that at these laboratory temperatures MWP substantially exceeds EWL. Unfortunately, we are unaware of appropriate data on simultaneous measurements of oxygen consumption and EWL for *Perognathus* spp., which would allow us to demonstrate directly this contention (none of the rodent species on which Figs. 6 and 7 are based are *Perognathus* spp.).

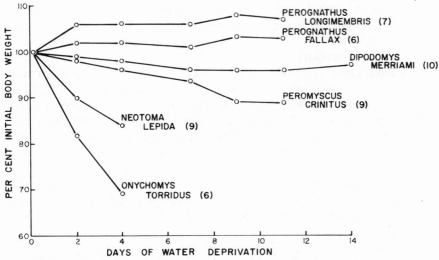

FIGURE 8 Response of body weight to water deprivation of rodents maintained on a diet of air-dried mixed birdseed (except *Neotoma lepida,* which was maintained on lab chow). The numbers represent sample size.

The gradient in the response of body weight to water deprivation for the cricetids is even more striking than in the heteromyids (Fig. 8). Prior to water deprivation the three species maintained or gained weight on a dry-food diet with water available. When water was deprived, while on the same diet, their weight responses differed markedly. *Peromyscus crinitus* at laboratory temperatures lost weight only slightly, while producing quite concentrated urine (Table 3), which suggests near equality of MWP and EWL. Abbott (1971) showed that *P. crinitus* can maintain weight on a dry-seed diet and produce very concentrated urine. Both *Neotoma lepida* (an herbivore) and *Onychomys torridus* (a carnivore) lose weight very rapidly (Fig. 8), while producing only moderately concentrated urine (Table 3). These data are similar to those of Lee (1963) and of Schmidt-Nielsen and Haines (1964) for these species, respectively, and suggest for each that MWP is considerably less than EWL, and that their limited abilities to concentrate urine exaggerate further the excess of water loss over MWP while on a dry diet.

The response of body weights to water deprivation (Fig. 8) while the mice were subsisting on a dry-seed diet at laboratory temperatures (21 to 25°C) suggests that, of the rodent species studied herein, the heteromyids should seldom, if ever, encounter nocturnal surface temperatures (see Fig. 1) at which MWP < EWL (see Fig. 7), provided seeds are available for metabolic water production. Further, among the cricetids, *Peromyscus crinitus* should have the option of subsisting on a granivorous diet with MWP > EWL, or shifting occasionally or permanently to a more succulent diet of animal or vegetable material. *Neotoma lepida* and *Onychomys torridus* appear to have no option but to be dependent at all times on an intake of succulent foods to meet their water requirements.

Thus, we predict that the granivorous heteromyids with a presumably highly favorable relationship between MWP–EWL and t_a would have urine concentrations in the field which reflect this relationship, and which reflect the annual fluctuations in nocturnal surface temperatures. For the cricetids *N. lepida* and *O. torridus,* with an apparently unfavorable relationship between MWP–EWL and t_a, we predict a reliance on succulent foods with urine concentration reflecting the osmotic loads imposed by those foods. And for the cricetid *Peromyscus crinitus,* with an apparently intermediate relationship between MWP–EWL and t_a, we predict considerable flexibility in meeting water needs, with urine concentrations reflecting this flexibility.

Seasonal fluctuations in concentration of the urine collected from these species in the field confirm very nicely these predictions, particularly for the most abundant species (see Figs. 2, 3, and 4). All three species of granivorous heteromyids show pronounced cycling of field-derived urine concentrations, with renal performance apparently geared to water elimination during the cold winter months and to water conservation during the warmer summer months. With summer monthly mean minimal air temperatures at the study sites never exceeding 24°C (see Fig. 1), it is doubtful that these species (*D. merriami, P. fallax,* and *P. longimembris*) are ever in a position where MWP < EWL, and therefore are never faced with negative water balance as long as foods (seeds) are available that provide a high yield of metabolic water. During the occasional, uncommonly hot summer night when surface temperatures might exceed those at which MWP > EWL, the animals need simply to stay in their underground burrows, subsisting on stored seeds while buffered from the less favorable surface conditions.

Field urine concentrations of the cricetids *Onychomys torridus* (Fig. 3) and *Neotoma lepida* (Fig. 4), although the data are rather meager, are quite stable seasonally and from year to year. Further, they appear to reflect very nicely the different osmotic loads imposed by the differing food habits of the two species: *O. torridus,* a carnivorous mouse, has uniformly very high urine concentrations, which very likely is a reflection of the high nitrogen concentration of the food (protein) and urine (urea); *N. lepida,* which relies heavily on green vegetation, has uniformly low urine concentrations, reflecting the succulent, low-proteinaceous nature of its diet.

As predicted, the cricetid *Peromyscus crinitus* has field urine fluctuations intermediate between those of the heteromyids and the other cricetids, reflecting its dietary flexibility in meeting its water needs (Fig.

2). Surprisingly, the midwinter urine concentrations of *P. crinitus* are very high, exceeding significantly those of the granivorous heteromyid *D. merriami* and comparable to those of the carnivorous cricetid *O. torridus;* this suggests the likelihood of a carnivorous diet (probably insects) for *P. crinitus* during the winter months. During the summer months, field urine concentrations of *P. crinitus* are maximal and very similar to those of *D. merriami*. A comparison of total osmotic pressure, urea, and chloride concentrations of field-collected urine of *P. crinitus* and *D. merriami* during 1971 at Joshua Tree (Fig. 9) shows the former to

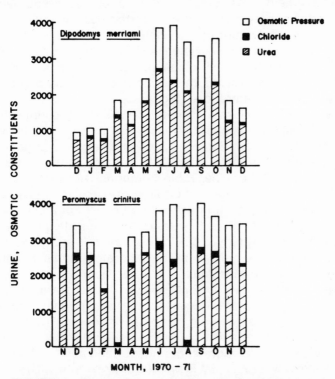

FIGURE 9 Some major osmotic constituents of the field-collected urine of *Dipodomys merriami* and *Peromyscus crinitus* from the Joshua Tree site during 1970 and 1971. The maximal height of each bar represents mean total osmotic pressure of the monthly sample, expressed as mOsm/liter; mean chlorinity is expressed in mEq/liter; mean urea concentrations are expressed in mmol/liter. Urea and chloride concentrations are designated by that portion of each bar bearing the appropriate shading. No chloride measurements were available for *D. merriami* in December 1970, and no urea measurements for *P. crinitus* in March and August 1971.

have very high concentrations of urea during the winter months, also indicative of a carnivorous diet. Although the urine data (Figs. 2 and 9) do not reveal differences during the summer months between *P. crinitus* and *D. merriami,* the former should be able to achieve the same state of positive water balance as the latter by either a carnivorous or granivorous diet, or by a combination of both.

Although we have categorized the rodents studied as granivorous (*D. merriami, P. longimembris,* and *P. fallax*), carnivorous (*O. torridus*), granivorous–carnivorous (*P. crinitus*), or herbivorous (*N. lepida*) largely on the basis of field urine samples, we must also look upon each as a food and water opportunist, with a preferred specialty but also taking advantage of unusual opportunities. Hence, we believe *P. longimembris* switched from a granivorous existence in early April 1973 to a carnivorous existence 1 week later with the emergence of larvae of the painted lady butterfly; accompanying this switch in diet was a pronounced increase in urine concentrations (Fig. 3).

Since field-derived plasma osmotic concentrations remained relatively stable while urine concentrations varied with seasonal changes in temperature, diet, or both (Figs. 2, 3, and 4), one would predict fluctuations in field-derived urine:plasma (U:P) osmotic ratios that would directly follow those of the urine. Table 4 demonstrates these fluctuations of U:P osmotic ratios for the Joshua Tree site between December 1970 and December 1971, and confirms this prediction. For those species for which comparable data are available from previous laboratory studies, the highest mean field-derived U:P ratios from this study are invariably higher than the means derived in the laboratory under conditions of water deprivation (see MacMillen, 1972, for laboratory values). As implied previously, these high U:P ratios are not necessarily indicative of superior capacities for water conservation through renal concentration for all species, but reflect varying combinations of renal water conservation and solute load imposed by the particular diet. Or, conversely, a species may be capable of greater concentrating abilities than it shows in the laboratory, where its diet is unnaturally restrictive. Thus, maximal field-collected urine concentrations for most of the species exceed laboratory-collected concentrations from water-deprived animals eating dry seeds, which suggests that urine concentrations reflect not only the amount of water in the diet, but also the nonaqueous constituents. One of the least explored and greatest remaining mysteries in the understanding of the water relations of desert rodents is the diet upon which each species subsists and the temporal changes in diet.

Sufficient recaptures of tritium-injected rodents for data on water turnover were obtained for only the heteromyids, *D. merriami* and *P. fallax,* at the Morongo Valley site (Fig. 5). These data reflect precisely the trends exhibited by the field urine data for these species (Figs. 2 and 4). As predicted by the relationship between MWP–EWL and t_a, turnover

TABLE 4 Mean urine: plasma osmotic pressure ratios of field-collected urine and blood samples from rodents at the Joshua Tree site during representative periods of 1970 and 1971

	Mean urine: plasma ± S.D. (sample size)				
Species	Dec. 1970	Mar. 1971	June 1971	Sept. 1971	Dec. 1971
Dipodomys merriami	2.7 ± 0.6 (16)	5.7 ± 1.7 (10)	11.4 ± 1.5, (10)	9.3 ± 3.1 (15)	4.8 ± 1.7 (2)
Perognathus fallax	2.5 (1)	5.4 (1)	9.3 ± 1.5 (2)	7.9 ± 2.7 (6)	—
Perognathus longimembris	—	4.4 ± 1.7 (8)	13.9 ± 0.9 (6)	3.7 ± 0.2 (2)	—
Neotoma lepida	3.7 ± 1.2 (2)	—	—	4.5 ± 0.9 (3)	—
Onychomys torridus	10.1 ± 0.4 (2)	—	8.4 (1)	10.3 (1)	9.3 (1)
Peromyscus crinitus	10.1 ± 0.9 (5)	7.7 ± 1.2 (3)	11.6 ± 1.2 (5)	12.5 ± 1.9 (9)	9.4 ± 3.3 (4)

of body water is greatest during the colder winter months when MWP greatly exceeds EWL, and is least during midsummer when MWP most closely approaches EWL. It is a pity that more winter turnover data are not available for the other species, as they, too, should be equally revealing.

Because, serendipitously, this study was conducted during a "normal" rainfall year (1971), followed by a year of extreme drought and virtually no annual plant germination (1972), followed by a moderate rainfall year with highly successful annual plant germination and maturation (1973), we are in a unique position to comment on the relationship between states of water balance as indicated by blood and urine osmotic properties and relative degrees of water availability in the environment (or its reciprocal, "water stress"). Measurements of urine and plasma osmotic pressure in all six species of rodents showed no discernible differences between nondrought years (1971 and 1973) and an intervening year of drought (1972). In addition, for the rodents which were found at both Joshua Tree and Morongo Valley, the highest urine concentrations (and, implicitly, the greatest "water stress") occurred most typically in Morongo Valley in 1973, the more mesic of the two sites and during a year of massive annual vegetative growth. Thus, we must conclude that our data reveal no physiological symptoms of water stress that correlate with environmental water paucity in the year of abnormally low rainfall (1972). We conclude further that for desert rodents, whether they be primarily granivorous (heteromyids), herbivorous (Neotoma lepida), carnivorous (Onychomys torridus), or more generalized (Peromyscus crinitus), food and water occur in the same package. As long as the package is available, so also is water (in varying proportions of preformed and metabolic water). In addition, this reliance upon a fairly discrete food–water package allows for maximal efficiency in resource allocation, and minimizes competition for common resources in multispecies populations. This efficiency is further enhanced by microhabitat preference differences, with three species (*D. merriami, P. longimembris,* and *O. torridus*) occurring primarily on the desert floor and three (*N. lepida, P. crinitus,* and *P. fallax*) on the granitic ridges. That unusually high urine concentrations did not occur in the rodents during the summer of 1972 (which the data indicate in spite of small sample sizes) suggests that the combined food–water package was available in sufficient supply for the individuals sampled, even during a severe drought.

Why, then, did the rodent population at the Joshua Tree site diminish almost to zero during the 1972 period of drought? We believe, as has been suggested by Beatley (1969) for a similar desert rodent population, that this was due to reproductive failure in the rodents caused by a failure of annual germination and growth and by natural mortality. We do not believe that the reproductive failure was, as Beatley (1969) suggests, due to an inadequate supply of dietary water. Reproduction typically commences in February and March (French et al., 1967), which is the period that coincides with very low nocturnal air temperatures (Fig. 1). According to our predictions and data, these conditions yield an excess of metabolic water (Fig. 7), relatively low urine concentrations for all species (Figs. 2, 3, and 4), and therefore most likely more than ample dietary water to provide for reproductive requirements. Rather, we support Beatley's (1969) alternative suggestion concerning the importance of annual vegetation, that "essential food derivatives (especially vitamins) in the fresh plant material could also be expected to play roles of importance in the physiology of reproduction."

Thus, our field studies of the water relations of two nocturnal desert rodent populations suggest that water stress, which is likely to occur only while active on the surface, is a seldom-encountered phenomenon. This suggestion is further supported by Kenagy's (1973) observation that *Dipodomys merriami* and *Perognathus longimembris* can meet their foraging requirements with as little as 1 hour per night of exposure to surface conditions. Although the several rodent species we studied appear to differ markedly in their potential relationships between MWP–EWL and t_a, those (*Neotoma lepida* and *Onychomys torridus*) in which MWP appears to be less than EWL during most nocturnal surface activity readily compensate for the excess water loss through the prudent choice of succulent foods. Those (*Dipodomys merriami, Perognathus longimembris,* and *P. fallax*) in which MWP appears to exceed EWL during virtually all surface activity (even including periods of drought) seem to be, indeed, independent of exogenous water while subsisting on seeds, although they do take in varying amounts of water in their foods. And one rodent (*Peromyscus crinitus*) appears able to exploit, and apparently does, either of the food and water packages depending upon its availability.

Acknowledgments

This study was supported by NSF Grant GB-17833 to R. E. MacMillen and NSF Grant GB-30132 and NWF fellowship award to E. A. Christopher. In addition, we gratefully acknowledge the arduous and voluntary efforts of K. D. Abbott, R. V. Baudinette, T. J. Case, D. E. Grubbs, T. E. Meehan, and L. C. Murdock. R. E. MacMillen is especially grateful to his parents, Mr. and Mrs. H. N. MacMillen, for use of their cottage at Joshua Tree. All of us also appreciate the many benefits provided by the Joshua Tree Early Bird Association.

LITERATURE CITED

Abbott, K. D. 1971. Water economy of the canyon mouse *Peromyscus crinitus stephensi*. Comp. Biochem. Physiol. 38A:37–52.

Beatley, J. C. 1969. Dependence of desert rodents on winter annuals and precipitation. Ecology 50:721–724.

Carpenter, R. E. 1966. A comparison of thermoregulation and water metabolism in the kangaroo rats *Dipodomys agilis* and *Dipodomys merriami*. Univ. Calif. Publ. Zool. 78:1–36.

French, N. R., B. G. Maza, and A. P. Aschwanden. 1967. Life spans of *Dipodomys* and *Perognathus* in the Mojave desert. J. Mammal. 48:537–548.

Holleman, D. F. 1972. Biological half-time and tracer experiments. Int. J. Appl. Radiat. Isotopes 23:341.

———, and R. A. Dieterich. 1973. Body water content and turnover in several species of rodents as evaluated by the tritiated water method. J. Mammal. 54:456–465.

Kenagy, G. J. 1973. Daily and seasonal patterns of activity and energetics in a heteromyid rodent community. Ecology 54:1201–1219.

Lee, A, K. 1963. The adaptations to arid environments in wood rats of the genus *Neotoma*. Univ. Calif. Publ. Zool. 64:57–96.

MacMillen, R. E. 1972. Water economy of nocturnal desert rodents. in G. M. O. Maloiy (ed.), Comparative physiology of desert animals. Symp. Zool. Soc. London 31:147–174.

———, and D. E. Grubbs. In press. The effects of temperature on water metabolism in rodents. In H. D. Johnson (ed.), Progress in Animal Biometeorology. Swets and Zeitlinger, Amsterdam.

Minnich, J. E., and V. H. Shoemaker. 1970. Diet, behavior and water turnover in the desert iguana, *Disposaurus dorsalis*. Amer. Midl. Nat. 84:496–509.

Raab, J. L., and K. Schmidt-Nielsen. 1972. Effect of running on water balance of the kangaroo rat. Amer. J. Physiol. 222:1230–1235.

Schmidt-Nielsen, K. 1972. How animals work. Cambridge University Press, New York. 114 p.

———, and H. B. Haines. 1964. Water balance in a carnivorous desert rodent, the grasshopper mouse. Physiol. Zool. 37:259–265.

PHOTOSYNTHETIC ADAPTATIONS TO HIGH TEMPERATURE

H. A. Mooney, O. Björkman, and J. Berry

Department of Biological Sciences, Stanford University, Stanford, California, and Department of Plant Biology, Carnegie Institution of Washington, Stanford, California

Abstract

Our study is focused on thermal responses and adaptations of plants that are native to cool coastal and hot desert habitats. In this paper we have drawn together information that illustrates different patterns of adaptation to the same habitat as a result of differing seasonal activity of plants, apparent adaptive modification of the characteristics of a single plant with seasonal changes in the climate, and evidence that indicates considerable direct control of plant development by temperature. There are corresponding differences among the species studied in the capacity to acquire water, photosynthesize, grow, and survive as a function of temperature. Several of the species studied possess the C_4-dicarboxylic acid pathway of photosynthesis. Differences between this pathway and the usual C_3 pathway are discussed in terms of adaptive potential. Detailed studies of how these potentials are utilized by these species are in progress. Results to date illustrate that plants with the C_4 pathway can have quite different responses to temperature, and that the potentials of this pathway may be utilized to somewhat different advantage by plants adapted to different environments or by plants adapted to the same environment in different ways.

INTRODUCTION

We are engaged in an experimental program to identify and understand the responses and adaptations of plants to contrasting thermal regimes. In this program we are making a special effort to bridge the gap that usually exists between investigations of field ecology, whole plant physiology, and biochemical or biophysical studies. Consequently, studies of the natural distribution, phenology and performance of plants in their native habitats are carried out in concert with studies utilizing transplant gardens in contrasting climatic regions and controlled growth facilities. These facilities permit simulation of native environments, investigation of the dependence of plant characteristics upon the environment, and serve as a source of plant material for laboratory studies that can be related to the native habitat. The correlation between field and laboratory studies is also being tested by use of our mobile laboratory (Björkman et al., 1973). Our efforts at a physiological level have focused on photosynthetic differences, particularly the significance of the C_4-dicarboxylic acid (C_4) pathway of photosynthesis. Our long-range objective is to understand and evaluate at least the most obvious components of these sets, both mechanistically and in terms of their significance to the plant in its natural environment.

Death Valley, which has one of the hottest climates on earth, is one of the central focus habitats of our study. The plants which grow there must tolerate temperatures that often exceed 50°C during the summer. These plants are thus of special interest when considering adaptations to high temperatures, since they represent the potential limits of adaptability for terrestrial plants.

The temperatures of Death Valley vary greatly during any given day as well as between seasons. The average daily temperature range during any month is between 14 and 17°C (Fig. 1). Average daily maxima are over 46°C in July and 18°C in December and January. The highest air temperature ever recorded in the valley is 56.7°C and the lowest −9.4°C. The record high soil surface temperature is 94°C (U.S. Weather Bureau Records for Death Valley).

Temperature extremes offer only one part of the environmental adaptive challenge for plants inhabiting Death Valley. An average of less than 50 mm of rain falls in the valley during the year, principally during the winter. The evaporative loss of this moisture is great because of the tremendous vapor pressure gradient (> 100 mbar during summer) that often exists between free water and atmosphere. Potential evapotranspiration is 4064 mm/year. This exceeds rainfall by 100-fold.

Any consideration of photosynthetic adaptation of Death Valley plants to temperature must therefore also consider adaptations to water limitations, since at times of highest temperatures water vapor pressure gradients are extreme. Furthermore, since diffusive pathways for carbon dioxide into the leaf and water vapor out of the leaf are the same, one would predict that plants which are photosynthetically active dur-

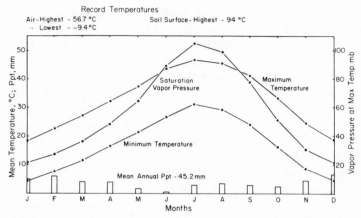

FIGURE 1 Mean daily maximum and minimum temperatures and monthly precipitation at Furnace Creek, Death Valley, California. Saturation vapor pressure of water at the maximum temperature is shown. The actual vapor pressure of water in the air seldom exceeds 15 mb during the year (unpublished records); thus the vapor pressure gradient for evaporation of water from a leaf varies from a few millibars in water to as much as 80 to 100 mb in summer.

ing the summer should have unusual mechanisms to gather and transport water and carbon dioxide.

The Death Valley environment contrasts sharply with the cool marine climate of our other study site at Bodega Head, located on an exposed coastal bluff adjacent to the Pacific Ocean. At this site there is little seasonal or diurnal temperature variation. Maximum daily temperatures in July are typically near 16°C and are accompanied by strong onshore winds and frequent ocean fog.

SEASONAL PLANT RESPONSE

Few woody perennial plant species inhabit the floor of Death Valley. The most prominent are *Prosopis juliflora*, a phreatophyte, *Larrea tridentata, Atriplex hymenelytra*, both xerophytes, and the subshrub, *Tidestromia oblongifolia*. These species are all found in close proximity at the base of the large alluvial fans above the playas. In these habitats, it is possible to examine the response of diverse plant forms to the extreme climatic conditions of the valley.

Although these species may exist side by side, their seasonal phenological and water stress patterns are quite dissimilar (Fig. 2). *Tidestromia* grows principally during the hot summer and flowers in autumn. Even though it transpires very rapidly in the middle of summer, it does not develop great water stress. During winter most of the

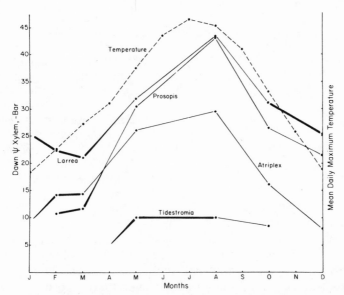

FIGURE 2 Seasonal course of dawn water potential of four woody plant species co-occurring in Death Valley. Values given are for a composite year for which the first measurements were made in August 1973 and the last in May 1974. Thickened portions of the lines indicate active growth periods.

aboveground portions of this subshrub are deciduous. *Prosopis,* a small deciduous tree, grows rapidly in late winter and spring. It keeps its leaves during summer, but even though it is a phreatophyte it develops high water stress. It is leafless during the winter months. *Larrea,* an evergreen shrub, grows during all but the hottest summer months, but flowers in spring. It has high water stress throughout the year, and maximum stress occurs in summer. *Atriplex hymenelytra* is also an evergreen shrub. It grows throughout the year, but most of its growth and flowering occur during winter. It does not develop quite such high water stress during summer as either *Larrea* or *Prosopis.*

Each of these species has apparently adapted to the same annual thermal regime in a dissimilar manner. All obviously tolerate the extremes of summer temperature and are photosynthetically active during summer. However, all except *Tidestromia* have much reduced activities during the summer months in comparison with late winter and spring, and only *Tidestromia* grows rapidly during summer.

There is a relationship between the period of most active growth of these species and their photosynthetic temperature response (Fig. 3). *Atriplex,* which grows during the coldest months, has a relatively low thermal optimum of photosynthesis; conversely, the summer-growing *Tidestromia* has an unusually high thermal optimum. It also appears that

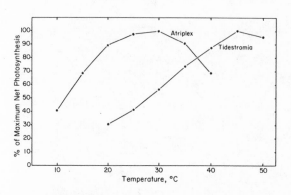

FIGURE 3 Temperature curves of photosynthesis of *Atriplex hymenelytra* and *Tidestromia oblongifolia* as determined on plants actively growing under natural conditions in Death Valley. For comparative purposes the curves are normalized; actual values are *Atriplex,* March, 100% = 7.6 μmol dm^{-2} min^{-1}; *Tidestromia,* July, 100% = 23 μmol dm^{-2} min^{-1}. (Data from Pearcy et al., 1971.)

the plants are photosynthetically active when they are under the least water stress. The capacity of *Tidestromia* to grow and photosynthesize during the highest temperature period coincides with the capacity to obtain and transport sufficient moisture even under the high evaporative demands of summer. In contrast, plants such as *Atriplex* exhibit high water stress during the summer period even though irrigated twice daily (Fig. 4). These seasonal changes of plant water potential could be due to temperature or indirectly to photoperiod. From transplant experiments in which *Tidestromia* and *Atriplex* were grown under similar photoperiods but different sets of other environmental variables, and from controlled environment growth experiments, it is clear that temperature is the main determinant of the responses of these species (Björkman et al., 1974a, 1974b). Similar studies of *Larrea* and *Prosopis* have not yet been conducted.

SEASONAL MODIFICATIONS

During the different seasons there is marked difference in the appearance of some of the plants growing in Death Valley. Similar differences are also evident in the same plants when grown in growth chambers at

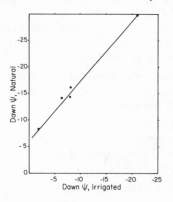

FIGURE 4 Correlation between the dawn xylem water potential of plants of *Atriplex hymenelytra* growing in Death Valley under natural conditions and under conditions of daily irrigation. Measurements encompass cool winter and hot summer conditions. (Data from Mooney et al., 1974.)

Figure 5 Plants of *Atriplex hymenelytra* growth under cool (16°C day; 10°C night) and hot (45°C day; 31°C night) thermal conditions. Note the difference in leaf size.

different temperatures (Fig. 5) (Mooney et al., 1974). In late winter (or at low growth temperature) the leaves of *Atriplex hymenelytra* are a dark green and succulent. During summer (or at high growth temperature) the leaves are less succulent, smaller, and very reflective. In summer, leaves that developed during the winter resemble new summer leaves except in size. These changes may have a considerable effect on the energy balance of the leaves during the summer when tissue temperatures might approach lethal levels. The decreased water content leads to a decreased absorptivity of infrared radiation. Increased reflectance of radiation is also correlated with changes in leaf moisture content. The lower the moisture content, the lower the reflectivity (Fig. 6). Leaves

FIGURE 6 Relationship between leaf water content of *Atriplex hymenelytra* and reflectivity at 500-nm radiation. (From Mooney et al., 1974.)

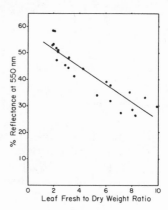

with a fresh to dry weight ratio of 2, as found on shrubs in Death Valley in the summer, reflect 50 to 60 percent of the incoming visible and infrared radiation. In contrast, leaves with higher moisture contents representing a fresh to dry weight ratio of 8, which are found on shrubs during late winter growing in Death Valley or on shrubs grown at cool temperatures, have lower reflectivity at all wavelengths (Fig. 7).

Electron microscope scans of leaf surfaces show that as the leaves become increasingly desiccated with increasing temperatures in Death Valley, the surface glands collapse. Salts (NaCl), which are evidently in solution in the hydrated winter leaves, crystallize and become trapped under the collapsed membranes in summer. Sodium chloride has a high reflectivity throughout the visible and infrared and probably contributes to the observed differences in reflectivity.

These changes might be interpreted as an indirect effect of leaf desiccation; however, the photosynthetic tissues of the leaves remain hydrated and active during the desiccation of the rest of the leaf. In addition, these changes appear to adapt the leaves to the high thermal load of summer by reducing the absorption of radiation, which in the absence of transpirational cooling would cause tissue temperatures to exceed air temperature.

Evidently, considerable modification of the photosynthetic characteristics of the plants themselves can occur during the growing season. Pearcy (unpublished data) has found that populations of *Atriplex lentiformis,* which grow on the California coast and in Death Valley, have virtually identical photosynthetic temperature responses when grown under the same thermal conditions (Fig. 8). However, plants of either population when grown under cool conditions have a much lower tem-

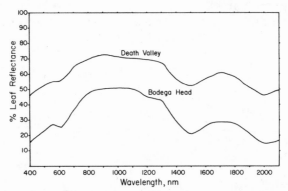

FIGURE 7 Spectral reflectance from leaves grown in Death Valley under natural conditions and at a garden located in a cool maritime climate (Bodega Head). Leaves were measured in August at both stations. (From Mooney et al., 1974.)

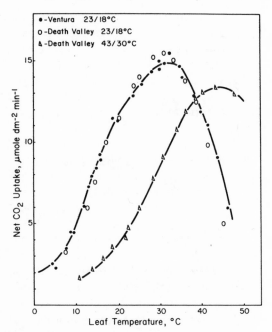

FIGURE 8 Comparison of the effect of temperature on the photosynthetic rate of plants of *Atriplex lentiformis* native to a cool coastal site (Ventura) and a desert site (Death Valley). Photosynthetic rates were determined in saturating white light in air of 0.03 percent CO_2. Growth conditions as in Figure 10 except light intensity was 220 to 250 × 10^3 ergs cm^{-2} s^{-1} (400 to 700 nm) and the temperatures were as indicated. (Pearcy, unpublished data.)

perature optimum of photosynthesis than do plants of the same population when grown under warm conditions.

The photosynthetic characteristics of *Larrea divaricata,* which is also capable of growth throughout most of the year (Oechel et al., 1972), may change with the season (Strain and Chase, 1966). However, not all plants are as flexible as these species. This is shown by experiments in which plants were transplanted to gardens in their native and contrasting environment.

GARDEN EXPERIMENTS

For these studies two gardens were established, one in Death Valley and one at Bodega Head, California. At both stations native plants from coastal and desert regions were planted at the first of April 1973 (Björkman et al., 1974a). By early summer at the desert garden all coastal species died (even though the plants were irrigated daily), except for coastal populations of *Distichlis spicata* and *Atriplex lentiformis,* which also occur as natives in Death Valley. Most native desert species were active in late spring. *Tidestromia oblongifolia,* however, showed maximum growth in summer. During the period of vigorous growth between May and June, mean daily maximum temperatures were 43°C.

Conversely, at Bodega the coastal endemics thrived (*Atriplex hastata, A. glabriuscula,* and *A. sabulosa*). The desert endemic *Tidestromia ob-*

longifolia, however, was totally unable to grow. *Atriplex hymenelytra,* another desert endemic, did grow and flower at Bodega, but only during the summer period in contrast to its winter and spring growth period at Death Valley.

It was clear that the plants which did best at either of the thermal extremes could not survive in the contrasting environment. This may be a general phenomenon.

PHYTOCELL EXPERIMENTS

Even though irrigation water was supplied twice daily to plants in the Death Valley garden, evaporative demands of 100 mbar were not uncommon and water stress must be considered in addition to high temperature stress. Experiments in naturally illuminated growth chambers (phytocells), where temperature and humidity could be controlled, were designed to separate these stresses (Björkman et al., 1973). For these experiments, the phytocells were programmed to simulate summer conditions at Bodega (16°C day; 11°C night) and Death Valley (45°C day; 31°C night). In both, humidity was kept as high as feasible, and the maximum vapor pressure deficit of the Death Valley phytocell was about one third of that which normally occurs under natural conditions. Nutrients and water were kept at nonlimiting levels, and radiation and turbulence conditions were equal. The growth of the two desert species, *Tidestromia oblongifolia* and *Atriplex hymenelytra,* and two coastal species, *Atriplex glabriuscula* and *A. sabulosa,* duplicated those found in the field (Björkman et al., 1974b) (Fig. 9). The two coastal species died in the high temperature regime but thrived in the cool regime. The desert species, *Tidestromia,* conversely showed no net growth in the cool temperatures but grew explosively under the hot regime. In fact, it had a mean daily growth rate of 26 percent and doubled its dry weight every 3 days! *Atriplex hymenelytra* grew in both chambers, and it was morphologically distinct in each. This points again to the exceptional flexibility of this species in contrast to the others studied. The failure of the coastal species at high temperature and of *Tidestromia* to grow at low temperature appears to be the result of direct effects of temperature on the metabolic capacity of these plants.

PHOTOSYNTHETIC PATHWAYS

We would now like to focus attention on the mechanisms by which these plants fix carbon dioxide and the extent to which differences be-

FIGURE 9 Plants of two coastal species (*Atriplex glabriuscula* and *A. sabulosa*) and two desert species (*A. hymenelytra* and *Tidestromia oblongifolia*) after 22 days growth under simulated coastal (16°C day; 11°C night) and desert (45°C day; 31°C night) thermal regimes.

tween them can be accounted for by such photosynthetic differences. Current evidence (Björkman, 1973) suggests that plants which have C_4 photosynthesis have evolved physiological specialization that permits the plant to extract CO_2 from the atmosphere by reaction with P-enolpyruvate carboxylase (C_4) rather than by the reaction with ribulose diphosphate carboxylase (C_3) used by other plants. Figure 10 condenses the results of many comparative studies on the two *Atriplex* species *A. rosea,* a C_4 species, and *A. patula,* a C_3 species. The curves depict the photosynthetic performance of these two plants in response to changes in the principal components of the physical environment. The responses of these two closely related plants are typical of the differences between C_3 and C_4 plants in general.

In general, C_4 species have higher maximum rates of photosynthesis (CO_2 uptake). This difference is most dramatic at high light intensity (full sunlight) and at warm temperature. If light intensity incident on the leaf falls much below 20 percent of full sunlight or if the temperature is below 20°C, there is little difference between the performance of these plants.

Increasing leaf temperature causes the potential for transpiration of water from the leaf to increase. If the rate of CO_2 uptake does not also increase, a decreased efficiency of water use must result (milligrams of CO_2 absorbed per grams of H_2O transpired). *Atriplex patula,* the C_3 plant, does not increase its photosynthetic rate with increased temperature above about 22 to 25°C, whereas the C_4 plant *Atriplex rosea* continues to increase to about 30°C. Thus, the water-use efficiency of *A. rosea* will not be decreased as much as that of *A. patula* by increasing the temperature from 20 to 30°C. This is a general phenomenon of most C_4 species. Some C_4 plants like *Tidestromia* do not reach temperature optimum until well above 40°C.

Another important point is the extent to which CO_2 uptake declines when the concentration of CO_2 within the leaf falls below atmospheric concentration. The C_4 species is clearly superior in this respect. This increased capacity to scavenge CO_2 means that C_4 plants can constrict their stomata, which in effect reduces the CO_2 concentration inside the leaf during photosynthesis, with less effect on CO_2 uptake than in a C_3 species. Constriction of the stomata will also affect the rate of transpiration, which will be the same for C_3 and C_4 plants. However, because of its greater capacity to extract CO_2, we may predict that the C_4 plants can maintain a higher rate of CO_2 uptake for any given stomatal resistance,

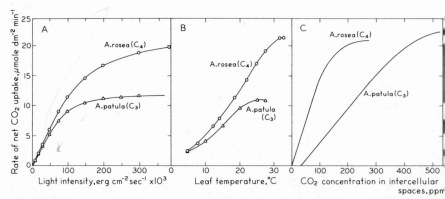

FIGURE 10 Comparative photosynthetic characteristics of the C_3 species *Atriplex patula* ssp. *hastata* and the related C_4 species *A. rosea* in air with 21 percent O_2. The plants were grown under identical controlled conditions of 25°C day and 20°C night, 110×10^3 ergs cm^{-2} s^{-1} (400 to 700 nm) light intensity for 16 h/day, and ample water and nutrient supply. Photosynthetic rates of single attached leaves are given as functions of (A) light intensity at 27°C leaf temperature, 21 percent O_2 and 0.03 percent CO_2, (B) leaf temperature at 250×10^3 ergs cm^{-2} s^{-1} and 0.03 percent CO_2, and (C) intercellular space CO_2 concentration at 250×10^3 ergs cm^{-2} s^{-1} and 27°C temperature. (From Björkman, 1973.)

or, conversely, that it would have a lower rate of transpiration at any given rate of CO_2 uptake. The C_4 species thus would be expected to have an increased water-use efficiency for any given photosynthetic rate (Fig. 11). This effect is in addition to the effect of temperature on water-use efficiency described earlier.

We should point out, however, that these comparisons are theoretical calculations; it is indeed difficult to extrapolate from them to the constantly changing conditions that prevail in nature. We are attempting to bridge this gap in our program by conducting our studies in the field and in the laboratory.

The Death Valley natives, *Tidestromia* and *Atriplex hymenelytra* are C_4 species; *Prosopis* and *Larrea* are C_3 species. The high thermal optimum for photosynthesis of *Tidestromia* is in part due to the presence of the C_4 pathway, which, coupled with a high potential for water-use efficiency, is obviously of adaptive advantage to *Tidestromia*, which confines its growth to the summer. Evidence (Pearcy et al., 1971) indicates that *A. hymenelytra* achieves a higher water-use efficiency than *Tidestromia* by having somewhat higher diffusion limitation of photosynthesis and by concentrating photosynthetic activity in the cooler portions of the year when evaporative demands are lower. *Larrea* and *Prosopis* also do this; perhaps this is related to their success in Death Valley even without C_4 photosynthesis.

It might be inferred that the failure of *Tidestromia* to grow at low

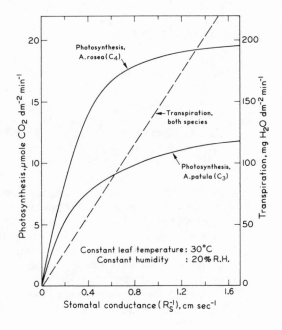

FIGURE 11 Rates of photosynthesis and transpiration for the C_3 species *Atriplex patula* and the C_4 species *A. rosea* as a function of stomatal conductance at a constant vapor-pressure deficit and a leaf temperature of 30°C. The rate of transpiration at any given stomatal conductance is identical for the two species. Rates of photosynthesis are calculated from the data presented in Figure 10. (From Björkman and Berry, 1973.)

temperature indicates that C_4 photosynthesis cannot function at low temperature. However, C_4 plants can potentially persist in cool climates. This can be seen by considering the results given earlier from the phytocells. The two coastal species, *Atriplex glabriuscula* and *A. sabulosa*, originate from similar cool maritime climates. As indicated, both have similar vigorous growth responses under cool growth conditions and neither survive the simulated hot Death Valley conditions. Each, however, possesses a different photosynthetic pathway; *A. sabulosa* has C_4 photosynthesis and *A. glabriuscula*, C_3.

At the same time it can be surmised that the capacity of *Tidestromia* to photosynthesize in such a remarkable manner under such high temperatures involves not only efficient photosynthetic use of water but also the capacity to acquire and transport the water under great evaporative demands. Furthermore, specialized features of the photosynthetic apparatus must exist, which enable it to maintain its integrity under these high temperatures. These problems are now the focus of our research efforts.

Acknowledgments

These studies were supported by NSF Grants GB-35854 and GB-35855.

LITERATURE CITED

Björkman, O. 1973. Comparative studies on photosynthesis in higher plants. Photophysiology VIII:1–63.

———, and J. Berry. 1973. High-efficiency photosynthesis. Sci. Amer. 225:80–93.

———, M. Nobs, J. Berry, H. Mooney, F. Nicholson, and B. Catanzaro. 1973. Physiological adaptation to diverse environments; approaches and facilities to study plant responses to contrasting thermal and water regimes. Carnegie Inst. Year Book 72:393–403.

———, M. Nobs, H. Mooney, J. Troughton, J. Berry, F. Nicholson, and W. Ward. 1974a. Growth responses of plants from habitats with contrasting thermal environments: transplant studies in the Death Valley and the Bodega Head experimental gardens. Carnegie Inst. Year Book 73:748–757.

———, B. Mahall, M. Nobs, W. Ward, F. Nicholson, and H. Mooney. 1974b. Growth responses of plants from habitats with contrasting thermal environments: an analysis of the temperature dependence of growth under controlled conditions. Carnegie Inst. Year Book. 73:757–767.

Mooney, H., O. Björkman, and J. Troughton. 1974. Seasonal changes in the leaf characteristics of the desert shrub *Atriplex hymenelytra*. Carnegie Inst. Year Book 73:846–852.

Oechel, W., B. Strain, and W. Odening. 1972. Photosynthetic rates of a desert shrub, *Larrea divaricata* Cav., under field conditions. Photosynthetica 6:183–188.

Pearcy, R., O. Björkman, A. Harrison, and H. Mooney. 1971. Photosynthetic performance of two desert species with C_4 photosynthesis in Death Valley, California. Carnegie Inst. Year Book 70:540–550.

Strain, B., and V. Chase. 1966. Effect of past and prevailing temperature on the carbom dioxide exchange capacities of some woody desert perennials. Ecology 47:1043–1045.

DROUGHT ADAPTATION IN CRASSULACEAN ACID METABOLISM PLANTS

Irwin P. Ting and Stan R. Szarek†

*Department of Biology and The Philip L. Boyd
Deep Canyon Desert Research Center, University
of California, Riverside, California*

Abstract

Biophysical and biochemical adaptations of succulent plants to arid environments are discussed. The ideal succulent form, which is spherical in shape, minimizes surface area per unit volume and increases boundary layer resistances (R_a). The spherical shape results in reduced water loss per unit area, but also decreases the surface available for CO_2 assimilation. Water loss is additionally reduced by high cuticular resistances (R_c). Strict control of stomatal resistances (R_s) also minimizes water loss and CO_2 assimilation. The mesophyll resistance (R_m) for CO_2 assimilation can be low throughout the night and approaches values reported for C_4 plants. Maximum CO_2 assimilation occurs with low R_m values.

The physical shape and gas transfer resistances coupled with nocturnal stomatal opening and crassulacean acid metabolism result in increased water-use efficiency by succulent plants. Transpiration ratio (TR) can be lower than in C_4 plants, ranging from 50 to 150 under ideal conditions. The estimated efficiency of carbon assimilation (E) approaches the 0.8 to 0.9 range of C_4 plants and exceeds the 0.2 to 0.3 range for C_3 plants. These adaptations increase the fitness of succulent plants in arid environments.

---

† Present address: Department of Botany and Microbiology, Arizona State University, Tempe, Arizona.

INTRODUCTION

There are many and varied plant adaptations to arid conditions. Some arid-zone plants avoid heat and water stress by growing only during favorable periods. Some exist only in washes and along watercourses, their deep roots penetrating the water table. Others have the capacity to withstand extreme tissue water stress while carrying out their necessary life processes. For ultimate growth and development in any environment, however, CO_2 must be assimilated.

The life of the green plant is centered around the process of photosynthesis. Photosynthesis, the assimilation of CO_2 and subsequent reduction to the level of carbohydrates using light energy, can be described in the form of a mass transfer equation (Penman and Schofield, 1951):

$$P = \frac{D' \, \Delta v'}{R_a + R_s + R_m} \tag{1}$$

where P = CO_2 assimilation rate in mg of CO_2 dm^{-2} h^{-1}
  D' = diffusion coefficient for CO_2
  $\Delta v'$ = CO_2 concentration difference between the atmosphere and leaf sink in mg of CO_2 cm^{-3}
  R_a = boundary layer resistance to diffusion in s cm^{-1}
  R_s = stomatal resistance to diffusion in s cm^{-1}
  R_m = mesophyll or residual resistance to CO_2 assimilation in s cm^{-1}

Assuming that $\Delta v'$ is constant, i.e., 59.3 mg of CO_2 $cm^{-3} \times 10^5$ at STP, and since D' is also a constant, it is the resistance terms of equation (1) that are most important in regulating CO_2 assimilation. The boundary layer resistance R_a is largely a function of the shape and size of the plant surface, as well as air movement. R_s is a variable resistance depending on stomatal aperture and is also a function of the stomatal spacing and thickness of guard cells. R_m is a residual term, which accounts for the transport of CO_2 to the assimilating center and the biochemistry associated with CO_2 fixation.

Similarly, transpiration or water loss can be described with a transfer equation:

$$T = \frac{D \, \Delta v}{R_a + R_s} \tag{2}$$

where T = transpiration rate in mg of H_2O dm^{-2} h^{-1}
  D = diffusion coefficient for H_2O
  Δv = H_2O vapor concentration difference between the leaf and atmosphere sink in mg of H_2O cm^{-3}
and R_a and R_s are the same as in equation (1).

An additional resistance that parallels R_s is the resistance to gas transfer directly through the interstomatal epidermal surface. This resistance, R_c, or cuticle resistance, is usually constant and is not as important as the other resistances in the control of gas transfer.

Although photosynthesis (P) is the most important biological process on earth, it is water that limits plant growth and productivity throughout the world. Even under mesic agricultural conditions, plants are apt to wilt during midday. Under arid desert conditions, water becomes paramount and clearly overshadows all other considerations. High temperatures, of course, are important but mostly insofar as heat effects water evaporation.

An important estimate, which is a measure of water-use efficiency in relation to CO_2 assimilation, it is the transpiration ratio (TR) (Black, 1971), which is the ratio T/P, or the ratio of equation (2) to equation (1):

$$\text{TR} = \frac{T}{P} = \frac{D \, \Delta v/R_a + R_s}{D' \, \Delta v'/R_a + R_s + R_m} \tag{3}$$

High TR values indicate high water loss relative to CO_2 assimilation and for many mesic plants may be in the range of 500 to 1000. Since the term Δv is a function of the evaporative demand of the environment, T/P varies with evaporative demand. For this reason we have attempted to define a term E, which is similar to TR but more of an intrisic plant factor (Ting et al., 1972). We define E as the ratio of actual photosynthesis [P of equation (1)] to the theoretical maximum photosynthesis (P_{max}) defined by the equation

$$P_{max} = \frac{D' \, \Delta v'}{R_a + R_s} \tag{4}$$

Hence, ignoring R_a, E is as follows:

$$E = \frac{R_s}{R_s + R_m} \tag{5}$$

E will vary from 0 to 1, the latter value being the theoretical maximum value, i.e., at $R_m = 0$.

This analytical discussion exemplifies a great paradox of plant biology. For photosynthesis and CO_2 assimilation to occur, stomates must be open, i.e., R_s must be low, but when this occurs great quantities of water are lost through the stomates by evaporation from the moist internal leaf tissues. Furthermore, CO_2 assimilation is a light-dependent process, and during the daylight hours temperatures are high and evaporation is greatest. It seems obvious that morphological, physiological, or biochemical means to reduce TR or increase E are adaptive for arid conditions. Means to alter TR or E reside largely in altering the resistances of equations (1) and (2).

Succulent Form

The ideal succulent form that minimizes surface area with respect to volume would be a sphere. Many barrel cacti and mammillarias of the southwest United States approach this form, but these cacti usually have an increased surface area because of raised tubercles and fluting. A comparison between the surface area of a perfect sphere of 1-liter volume with a leaf-like disk 1 mm thick, but still 1 liter in volume, exemplifies the adaptive nature of the highly succulent form. The sphere, or model succulent, would have a radius of about 6.2 cm and a surface area of 484 cm². The leaf-like disk of the same volume would have a radius of 56.5 cm and a total surface area of over 20,000 cm², or 41 times greater. Assuming total CO_2 assimilation and water loss as a function of the surface area, the leaf-like disk is clearly adapted to maximizing photosynthetic surface area, but as a consequence maximizes water loss. The succulent form, contrarily, minimizes water loss, but also minimizes the surface for CO_2 assimilation.

Powell (1940) developed equations to describe the evaporation from disks and spheres, which can be used to compare the succulent form with the leaf-like form. His equations are as follows:

$$EVAP_{sphere} = 2.1 \times 10^{-7} \, (VD)^{0.59} D^{-1} \qquad (6)$$

$$EVAP_{disk} = 3.3 \times 10^{-7} \, (VD)^{0.65} D^{-1} \qquad (7)$$

where $EVAP$ = evaporation, g cm^{-2} s^{-1} mm Hg^{-1}
V = air movement, cm s^{-1}
D = diameter, cm

It can be seen that, per unit area, a disk will lose more water than a sphere with the same diameter. Hence, the total evaporation from a disk-like surface will always be greater than from a spherical surface of the same area (Fig. 1). Further, a disk has more effective surface for CO_2 assimilation than a sphere.

An additional complication of the succulent form, due to its greater mass, is a higher heat capacity. Hence, succulents heat to a greater extent than nonsucculents, although more slowly (Ansari and Loomis, 1959); yet succulents can withstand great heat. MacDougal (1921) reported *Opuntia* growth between 9 and 50°C. The two most obvious detrimental features of higher heat capacity are the greater probability of protein denaturation and excessive water loss. The high mucilage content of some cacti (Spoehr, 1919) may protect protein denaturation (Henckel, 1964). With respect to water loss the problem appears particularly acute, because evaporation of water is one mechanism to dissipate heat (Gates, 1962). Adaptations to reduce water loss will result in even higher temperatures.

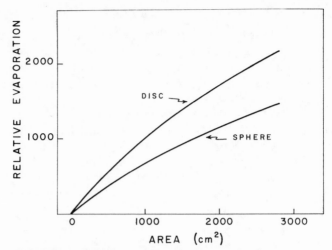

FIGURE 1 Relative evaporation per unit of vapor pressure of disk compared to a sphere with equal area. Calculations from the equations of Powell (1940) using unit air flow.

ADAPTATIONS TO REDUCE WATER LOSS

Physical

Although incipient drying of mesophyll evaporative surfaces is possible (Whiteman and Koller, 1964), variations in transpiration rates are essentially a function of resistances to gas transfer. We know of no careful estimates of the boundary layer resistance R_a, for succulent plants, but expect R_a to be greater for succulents than nonsucculents.

Calculations from equations (6) and (7) indicate that R_a for a sphere is greater than for a disk. As the diameter approaches zero, the boundary layer resistances of the two forms approach the same value. As the diameter increases, the ratio R_a (sphere) to R_a (disk) steadily increases (Fig. 2). At unit air velocity, the ratio is 2 at a diameter of about 45 cm. Hence, we predict that the boundary layer resistance, R_a, for the succulent form can be nearly double that of the disk-like leaf form. Furthermore, very spinous forms such as cacti would have a theoretically higher R_a than nonspinous forms.

The cuticular resistance, R_c, has been estimated for three *Opuntia* species (Ting et al., 1973). Estimates for *O. basilaris, O. acanthocarpa,* and *O. bigelovii* were 619, 624, and 1022 s cm⁻¹, respectively. These extremely high resistances are accounted for by a thick cuticle and heavy depositions of wax on the surface of older tissue (Fig. 3).

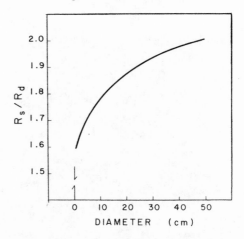

FIGURE 2 Ratio of boundary layer resistance (R_a) for a sphere (R_s) to a disk (R_d) as a function of diameter of the sphere or disk (see Fig. 1).

Stomatal resistance of succulent plants has been studied by a variety of workers. The general dimensions of stomates of many succulent plants are not much different from nonsucculents, although guard cells may be somewhat thicker and substomatal cavities deeper. This tends to increase the diffusion path length and results in higher minimum R_s estimates. Stomates of the Cactaceae and Crassulaceae that we have studied are not markedly sunken, but *O. basilaris* at least has a deep substomatal chamber extending from the epidermis through the several cell layers of hypodermis to the chlorenchyma (Fig. 3). Most mesic plants average 10,000 stomates/cm^{-2} or more (Verduin, 1950), but succulents as a group tend to have far fewer, averaging about 2,500/cm^2 (Ting et al., 1972).

Estimates of R_s for succulent plants are typically about 10 s cm^{-1} with minima of about 2 s cm^{-1}. These estimates can be contrasted with more mesic, nonsucculent plants, which have minima of about 1 s cm^{-1} (Table 1).

As well as generally higher R_s values, perhaps the most interesting feature of succulent plant stomates is their general reverse phase opening in comparison with nonsucculents. In general, succulent stomates open in the dark and close in the light (Nishida, 1963; Ting et al., 1967; Neales et al., 1968). When specimens of *Kalanchoe blossfeldiana* were preconditioned with cool temperatures and little diurnal temperature fluctuation, stomates tended to be open in light and closed in dark. If, however, plants were kept on a regime of high day temperatures and large diurnal temperature fluctuations, stomates tended to be open at night and closed during the day (Ting et al., 1967). Reverse phase stomatal opening of succulent plants seems to be the rule under natural conditions; however, day opening may occur during favorable moisture periods. At the Deep Canyon Desert Research Center, cacti with water

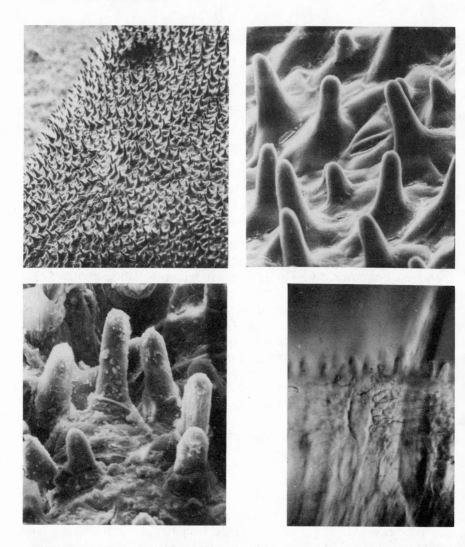

FIGURE 3 A. Scanning electron micrograph of surface of *Opuntia basilaris,* ×200. (Photo by Michael Adams.) B. Scanning electron micrograph of surface of young *O. basilaris* stem showing stomate and epidermal hairs, ×1750. (Photo by Michael Adams.) C. Scanning electron micrograph of old *O. basilaris* stem showing heavy wax-like deposit nearly obscuring stomate. (Photo by Michael Adams.) D. Light microscope photograph of *O. basilaris* stem section showing epidermis, multilayered hypodermis, and deep substomatal cavity extending to chloroenchyma. (Photo by C. A. Beasley.)

TABLE 1 Estimates of various gas-exchange parameters for crassulacean acid metabolism (CAM) plants and comparison with estimates for C_3 and C_4 plants

	CAM Dark	CAM Light	C_3	C_4
P_s(max)[a]	10–15	3–5	20–40	30–60
R_s(min)[b]	<3	>6	<1.0	>1.0
R_m(min)[b]	1–2	20–40	3.0	1.5
TR(ave)	25–100	150–600	600	300
E(max)	0.8–0.9	0.2–0.3	0.2–0.3	0.8–0.9

[a] mg of CO_2 dm^{-2} h^{-1}.
[b] s cm^{-1}.

potentials of −12 to −16 bars have stomates closed almost continuously. After precipitation and water uptake, stomates open at night and remain open during the first few hours of daylight (Szarek, 1974).

The generally higher R_a, R_c, and R_s values for succulent plants are clear adaptations to water stress conditions and regulation of gas exchange. An added feature of the high resistances is that there is greater stomatal control of gas exchange over the entire range of stomatal opening (Fig. 4). Nonsucculent mesic plants with low R_s values have near-maximum gas exchange after 50 percent of maximum stomatal opening. This is because R_a, which ranges from 0.5 to 2.0 s cm^{-1}, becomes the dominant resistance at low R_s, and subsequent reductions in R_s by stomatal opening do not affect the total resistance, i.e., $R_a + R_s$. Thus, stomatal regulation of gas exchange is most important at small stomatal apertures. Succulents, however, with minimum R_s values significantly greater than R_a, have stomatal regulation of gas exchange over the entire range of stomatal apertures.

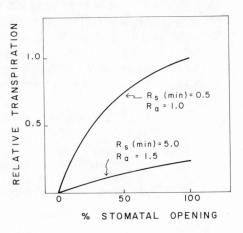

FIGURE 4 Transpiration on a relative basis for a disk-shaped leaf with a minimum R_s of 0.5 s cm^{-1} and assumed R_a of 1.0 s cm^{-1} in comparison with a succulent-shaped leaf or stem with a minimum R_s of 5.0 s cm^{-1} and assumed R_a of 1.5 s cm^{-1}.

These generally high gas-transfer resistances, coupled with the suc
culent morphology and nocturnal stomatal opening when the evapora
tive demand is low, result in significant water conservation adaptations

Biochemical

The adaptations minimizing water loss consequently result i
minimizing CO_2 uptake. To overcome this apparent anomaly, succuler
plants have evolved a biochemical mechanism by which CO_2 is obtaine
in the dark when stomates are open and the evaporative demand is low
The CO_2 is stored as malic acid and then released by decarboxylatio
during the subsequent light period when stomates are closed an
exogenous CO_2 cannot enter. This biochemical process, termed cras
sulacean acid metabolism (CAM), results in a massive diurnal fluctua
tion of malic acid (100–200 μEq. g^{-1} fresh weight).

We have established the following criteria to define CAM in a species
1. Low R_s or open stomates at night and high R_s or closed stomate
   during daylight.
2. Massive dark CO_2 fixation via the nonautotrophic carboxylatin
   pathway:

$$CO_2 + PEP \rightarrow OAA \rightarrow MAL$$

3. A large diurnal fluctuation of organic acids, viz., malic acid, at th
   expense of storage carbohydrates.
4. The main CAM metabolic events occurring in large, chloroplas
   containing cells.

Most CAM plants are distinctly succulent, such as those in the Cac
taceae, Crassulaceae, Euphorbiaceae, and Asclepiadaceae, but may b
quite coriaceous, as in the epiphytic Bromeliaceae and Orchidaceae.

Families known to have some CAM species are listed in Table 2
Twelve of these were previously cited (Ting et al., 1972) and the remair
der were designated from unpublished experiments conducted at th
Botanic Garden, University of California, Berkeley.

The presently envisioned steady flow of carbon through the CAI
pathway is depicted in Fig. 5 (Black, 1973; Ting, 1971). In the dark whe
stomates are open, stored carbohydrates are metabolized to solubl

TABLE 2 Families known to have CAM species

Agavaceae	Crassulaceae	Liliaceae
Aizoaceae	Cucurbitaceae	Orchidaceae
Asclepiadaceae	Didieraceae	Oxalidaceae
Asteraceae	Euphorbiaceae	Piperaceae
Bromeliaceae	Geraniaceae	Portulacaceae
Cactaceae	Labiatae	Vitaceae

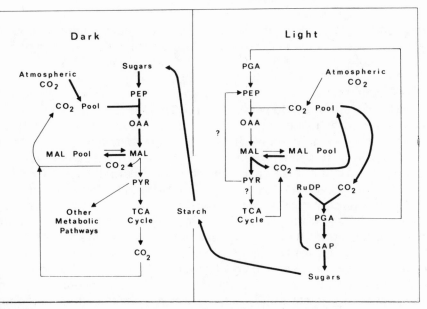

IGURE 5 Presently envisioned flow of carbon through the CAM photosynthe-
c pathway.

arbohydrates and, via glycolysis, ultimately to P-enolpyruvate (PEP).
arboxylation of PEP by PEP carboxylase (EC 4.1.1.31) forms oxaloace-
te (OAA), which is rapidly reduced to malate by a soluble malate
ehydrogenase (EC 1.1.1.37). A limited amount of aspartate is formed
om the oxaloacetate. The malate is stored as malic acid, probably in
acuoles. A certain proportion of the malate is metabolized, but this is
robably minimal compared to the amount accumulated (Ting and
ugger, 1968).

During the subsequent light period, when the stomates are closed,
e stored malic acid is decarboxylated. The switching mechanism from
ark carboxylation to light decarboxylation is poorly understood, but
vailable data indicate that the initial rate of dark malate synthesis is
wer than the initial rate of light malate decarboxylation. It is assumed
at malate decarboxylation is via the malic enzyme (EC 1.1.1.40) to
rm CO_2 and pyruvate. In some CAM plants, however, there is evidence
at PEP carboxykinase (EC 4.1.1.32) catalyzes the decarboxylation to
ive CO_2 plus PEP (Black, 1973).

Since the stomates are closed, the CO_2 released from malate is effec-
vely trapped in the tissue and is refixed by the photosynthetic pathway
utlined by Calvin and his associates (1951). Here the carboxylation is
y ribulose diphosphate (RudP) carboxylase (EC 4.1.1.39) to form two
hosphoglycerates (PGA). The PGA is reduced to glyceraldehydephos-

phate (GAP), the first carbohydrate of photosynthesis. The GAP ca
participate in a variety of metabolic reactions, but much is certainl
stored as starch. This pathway from RudP to the net production of GAI
is referred to as C_3 photosynthesis, because the first stable inte
mediate, PGA, is a 3-carbon compound.

The actual fate of the 3-carbon fragment resulting from malic aci
decarboxylation is not fully understood. There is too large a quantity t
be consumed by respiration, and it is suspected that most is converte
directly to starch or other storage carbohydrates by a reversal c
glycolysis (Ranson and Thomas, 1960). If PEP is the 3-carbon fragmer
of decarboxylation, the conversion directly to carbohydrates is plausi
ble. If, however, pyruvate is the 3-carbon fragment, then it is difficult t
visualize overcoming the energy barrier going from pyruvate to PEF
The enzyme, pyruvate-P_i dikinase, has been shown in CAM succulent
and could account for the conversion (Kluge and Osmond, 1971).

In any case, it is envisioned that the steady flow of carbon in the CAN
pathway results from a carboxylation of PEP to form malic acid whicl
accumulates. Concomitantly, stored carbohydrates decrease. Durin
the subsequent light period, malate decreases by decarboxylation an
starch is resynthesized by the usual C_3 photosynthetic pathway.

Basically, the CAM photosynthetic pathway differs from usual C
photosynthesis in that malic acid acts as a CO_2 storage compound. CO
is taken up and stored in malic acid at night. During the next day it i
released and assimilated by the usual C_3 photosynthetic pathway
Exogenous CO_2 cannot be assimilated during the day because of close
stomates. If, however, CO_2 is fed to cut tissue sections at the beginnin
of the light period when malate is high, sugars are preferentially synthe
sized (Kluge, 1971). On the other hand, if CO_2 is supplied at the end c
the light period when malic acid is low, malate is preferentially formec
Hence, it seems reasonable to conclude that malate is not a mandator
intermediate in CAM plant photosynthesis, but is an adaptive feature t
allow uptake of CO_2 at night when the evaporative demand is lowe

A recent observation from several laboratories (e.g., Osmond et al
1973) has suggested that by changing the photo-, thermo-, and/or hy
droperiod, the photosynthetic metabolism in some CAM plants shift
from the CAM mode, which couples malate synthesis to C_3 photosyn
thesis, to the C_3 mode independent of malate synthesis. Under thes
conditions, stomates are open during the day and CO_2 is assimilate
directly through RudP carboxylase. In distinctly arid zone CAM plant
such as cacti, water stress conditions most likely bring about the shif
from the C_3 mode to the CAM mode. In this sense, the CAM metaboli
pathway is truly opportunistic and highly adaptable to xeric environ
ments. While in the CAM mode the plants are largely only being sus

ained and, because of the limited amount of CO_2 assimilation possible, growth and productivity are predictably low. A shift to the C_3 mode under favorable moisture conditions would allow for greater CO_2 assimilation and hence the probability of greater growth and productivity.

It is worth comparing the metabolic similarities of C_4 photosynthesis and CAM. In C_4 photosynthesis (Hatch and Slack, 1970), malate appears as a mandatory intermediate in the coupling of PEP carboxylation of RudP carboxylation. Here the PEP carboxylation shown in Fig. 3 occurs in the light and not dark, and the immediate source of PEP is from the C_3 fragment that results from the decarboxylation of malate or other C_4 acid. The CO_2 released by malate decarboxylation is used directly in the RudP carboxylase reaction, and GAP is synthesized by the C_3 photosynthetic pathway. The PEP carboxylase reaction takes place in chloroplast containing relatively undifferentiated mesophyll cells, whereas the malate decarboxylation and assimilation by the C_3 pathway occurs in specialized, chloroplast-containing bundle sheath cells. Hence, like CAM, C_4 photosynthesis is a CO_2 concentrating mechanism. The main differences are the immediate source of PEP for the initial carboxylations and the separation of carboxylases spatially as in C_4 photosynthesis and temporally as in CAM.

A final point that emphasizes the reality of CAM is the isotopic fractionation between $^{13}CO_2$ and $^{12}CO_2$. C_3 photosynthetic plants, because of RudP carboxylase (Whelan et al., 1973), discriminate more against $^{13}CO_2$ than CAM or C_4 plants. Both use PEP carboxylase as the initial carboxylating protein. The mode discrimination (relative to a standard) for C_4 plants is about $-12°/oo$, and for C_3 plants, about $-28°/oo$ (Lerman et al., 1974). An estimate for *Opuntia basilaris* which always appears to be in the CAM mode at the Deep Canyon Research Center is $-13°/oo$ (data supplied by J. C. Lerman).

R_m

Estimations from the literature indicate that minimum R_m values for C_3 plants are about 3.0 s cm^{-2} and minima for C_4 plants are in the range of 0.5 to 1.5 s cm^{-1} (Table 1). The lower estimates for C_4 plants are accounted for by the PEP carboxylase CO_2 concentrating mechanism and apparent reduced photorespiration (Black, 1973). The few R_m estimations of CAM plants made are extremely variable, being a function of the CO_2 exchange mode during measurement. Our estimates of R_m for *O. basilaris* range from a minimum value of about 5 s cm^{-1} to values over 1000 s cm^{-1} (Szarek and Ting, 1974). This seasonal study of CAM indicates that during favorable moisture periods nocturnal R_m values are low, approaching 3 to 5 s cm^{-1}, and CO_2 uptake rates are high.

However, when water stress conditions are present, exogenous CO_2 assimilation is low and R_m values are high (Szarek and Ting, 1974). Furthermore, in *O. basilaris, R_m* varies diurnally (Szarek, 1974).

CO_2 Recycling

A unique adaptation for carbon conservation in succulent plants is their capability of recycling endogenously produced CO_2 through the CAM pathway (Szarek et al., 1973). During water stress periods when tissue water potentials are -12 to -16 bars, R_s values are high day and night with little CO_2 assimilation or water loss. Despite no exogenous CO_2 uptake, a measurable diurnal fluctuation of acidity indicates a recycling of endogenous CO_2 (Fig. 6). It appears that these plants can withstand severe and prolonged drought by hermetically sealing and minimizing water loss, while still reassimilating respiratory CO_2. Clearly, no growth could take place, but the plants could be sustained for extended periods. Immediately following precipitation, tissue water potentials increase to -2 or -3 bars, R_s and R_m decrease, exogenous CO_2 assimilation begins, and the diurnal fluctuation of acidity increases several fold (Fig. 6). If conditions are especially favorable, R_s may remain low during much of the day (Szarek, 1974).

TRANSPIRATION RATIO AND *E*

Previous estimations based on the literature indicate that TR values for C_3 plants are in the range of 600, and for C_4 plants, about 300 (Table 1). Although extremely variable, the estimates for C_4 plants are usually less than those for C_3 plants. Estimates for CAM plants are frequently less than C_4 plants and in the range of 50 to 150 (Ting et al., 1972).

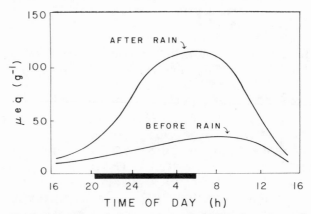

FIGURE 6 Diurnal fluctuation of acidity, largely malic acid, during drought and after precipitation.

Based on available literature, our calculations of E using equation (5) are about 0.2 to 0.3 for C_3 plants and as high as 0.8 to 0.9 for C_4 plants (Table 1). Estimations of E and *O. basilaris* are variable but approach 0.8 to 0.9 under ideal CO_2 assimilation conditions (Szarek and Ting, 1974).

GROWTH AND PRODUCTIVITY

We previously estimated that productivity of the photosynthetic groups in their own environments would be $C_4 > C_3 > $ CAM. Data reported by Whittaker (1970) indicate that desert scrub vegetation yields on the average 0.12 ton dry matter ha^{-1} yr^{-1}. Good agricultural land would yield about 1.16 tons ha^{-1} yr^{-1}, and lush tropical forest, about 3.6 tons ha $^{-1}$ yr^{-1}. We previously estimated agriculturally grown, uncultivated *Opuntia* at about 1.15 and cultivated at 4.16 tons ha^{-1} yr^{-1} (Ting et al., 1972). Pineapple fruit, assuming 20 percent dry weight, yielded about 1.6 tons ha^{-1} yr^{-1} with a density of 11,700 plants ha^{-1} (Wee, 1969). In the natural desert conditions at Deep Canyon, estimates of annual stem growth for three *Opuntia* species are, as expected, considerably below those cited here (Ting et al., 1973).

SUMMARY

Succulent plants are physically and biochemically adapted to maintain growth in arid regions. Their morphology, which varies from a disk-shaped, leaf-like form toward a spherical shape, minimizes water loss per total biomass. The increased boundary layer resistance, R_a, associated with spherical shapes reduces water loss per unit area, as does the greater average stomatal resistance (R_s) and the high cuticular resistance (R_c) attributed to the impervious cuticle. These increased resistances to water loss, coupled with a reverse phase stomatal opening, result in minimum water losses under arid conditions with high evaporative demands. These same adaptations, however, create increased heat loads and reduce CO_2 assimilation capacity. The latter, when coupled with a large residual resistance to CO_2 assimilation, R_m, further reduces the CO_2 assimilation potential. However, the E ratio can be quite high in succulents, approaching 0.9 under ideal conditions. Hence, these plants minimize long-term water loss, although being capable of maximum short-term CO_2 assimilation, and thus are well adapted for arid conditions.

The biochemical adaptation of crassulacean acid metabolism allows nocturnal CO_2 storage in the form of malic acid. Hence, exogenous CO_2 is taken up at night when the stomates are open and the evaporative demand is low. During the subsequent day, when stomates are closed and water loss is prevented, the decarboxylation of malic acid generates

CO_2 for photosynthesis without water loss. Therefore, this biochemical adaptation also tends to minimize water loss relative to CO_2 assimilation.

Acknowledgments

Research supported by a grant from the National Science Foundation administered through the US/International Biological Program, Desert Biome, Utah State University, Logan, and the University of California Chancellor's Patent Fund. We also thank Hyrum Johnson for much interest and stimulating discussion, and Mike Adams and Terry Yonkers for their contributions.

LITERATURE CITED

Ansari, A. Q., and W. E. Loomis. 1959. Leaf temperatures. Amer. J. Bot. 46:715–717.

Black, C. C. 1971. Ecological implications of dividing plants into groups with distinct photosynthetic production capacities. Adv. Ecol. Res. 7:87–114.

———. 1973 Photosynthetic carbon fixation in relation to net CO_2 uptake. Annu. Rev. Plant Physiol. 24:253–286.

Calvin, M., J. A. Bassham, A. A. Benson, V. H. Lynch, C. Duellet, L. Shou, W. Stepka, and N. E. Tolbert. 1951. Carbon dioxide assimilation in plants. Symp. Soc. Expt. Biol. 5:284–305.

Gates, D. 1952. Energy exchange in the biosphere. Biological Monographs. Harper & Row, New York. 151 p.

Hatch, M. D., and C. R. Slack. 1970. Photosynthetic CO_2-fixation pathways. Ann. Rev. Plant Physiol. 21:141–162.

Henckel, P. A. 1964. Physiology of plants under drought. Ann. Rev. Plant Physiol. 15:363–386.

Kluge, M. 1971. Studies on CO_2 fixation by succulent plants in the light. In M. D. Hatch, C. B. Osmond, and R. O. Slatyer (eds.), Photosynthesis and photorespiration, pp. 283–287. Wiley–Interscience, New York.

———, and C. B. Osmond. 1971. Pyruvate P_i dikinase in crassulacean acid metabolism. Naturwissenschaften. 58:414–415.

Lerman, J. C., E. Deleens, A. Nato, and A. Moyse. 1974. Variation in the carbon isotope composition of a plant with crassulacean acid metabolism. Plant Physiol. 53:581–584.

MacDougal, D. T. 1921. A new high temperature record for growth. Science 53:370–372.

Neales, T. F., A. A. Patterson, and V. J. Hartney. 1968. Physiological adaptation to drought in carbon assimilation and water loss of xerophytes. Nature 219:469–472.

Nishida, K. 1963. Studies on stomatal movement of crassulacean plants in relation to the acid metabolism. Physiol. Plant. 16:281–298.

Osmond, C. B., W.G. Allaway, B. G. Sutton, J. H. Troughton, O. Queiroz, U. Luttge, and K. Winter. 1973. Carbon isotope discrimination in photosynthesis in CAM plants. Nature 246:41–42.

Penman, H. L., and R. K. Schofield. 1951. Some physical aspects of assimilation and transpiration. Symp. Soc. Exp. Biol. 5:115–129.

Powell, R. W. 1940. Further experiments on the evaporation of water from saturated surfaces. Trans. Inst. Chem. Eng. 13:175–198.

Ranson, S. L., and M. Thomas. 1960. Crassulacean acid metabolism. Annu. Rev. Plant Physiol. 11:81–110.

Spoehr. H. A. 1919. The carbohydrate economy of cacti. Carnegie Inst. Wash. Publ. 287, 79 p.

Szarek, S. R. 1974. Physiological mechanisms of drought adaptation in *Opuntia basilaris* Engelm. & Bigel. Ph.D. Dissertation. University of California, Riverside, California.

———, H. B. Johnson, and I. P. Ting. 1973. Drought adaptation in *Opuntia basilaris*. Significance of recycling carbon through crassulacean acid metabolism. Plant Physiol. 53:539–541.

———, and I. P. Ting. 1974. Seasonal patterns of acid metabolism and gas exchange in *Opuntia basilaris*. Plant Physiol. 54:76–81.

Ting, I. P. 1971. Nonautotrophic CO_2 fixation and crassulacean acid metabolism. In M. D. Hatch, C. B. Osmond, and R. O. Slatyer (eds.), Photosynthesis and photorespiration, pp. 169–185. Wiley-Interscience, New York.

———, M. L. Dean Thompson, and W. M. Dugger. 1967. Leaf resistance to water vapor transfer in succulent plants: Effect of thermoperiod. Amer. J. Bot. 54:245–251.

———, and M. W. Dugger, Jr. 1968. Non-autotrophic carbon dioxide metabolism in cacti. Bot. Gaz. 129:1–5.

———, H. B. Johnson, and S. R. Szarek. 1972. Net CO_2 fixation in crassulacean acid metabolism plants. In C. C. Black (ed.), Net carbon dioxide assimilation in higher plants. Proc. Symp. South Sect. Am. Soc. Plant Physiologists, pp. 26–53. Cotton Inc., Raleigh, N.C.

———, H. B. Johnson, T. A. Yonkers, and S. R. Szarek. 1973. Gas Exchange and Productivity of *Opuntia* spp. U.S. International Biological Program Progress Report for 1972 (Desert Biome) RM 73-12. 22 p.

Verduin, J. 1950. Diffusion through multiperforate septa. In J. Franck and W. E. Loomis (eds.), Photosynthesis in plants, pp. 95–112. Iowa State College Press, Ames, Iowa.

Wee, Y. C. 1969. Planting density trials with Ananas comosus (L.) Merr. var. Singapore Spanish. Malay Agr. J. 47:164–174.

Whelan ., T., W. M. Sackett, and C. R. Benedict. 1973. Enzymatic fractionation of carbon isotopes by phosphoenol-pyruvate carboxylase from C_4 plants. Plant Physiol. 51:1051–1054.

Whiteman, P. C., and D. Koller. 1964. Saturation deficit of the mesophyll evaporating surfaces in a desert holophyte. Science 146:1320–1321.

Whittaker, R. H. 1970. Communities and ecosystems. Macmillan, New York. 158 p.

ENZYMIC MECHANISMS OF EURYTHERMALITY IN DESERT AND ESTUARINE FISHES: GENETICS AND KINETICS

George N. Somero

Scripps Institution of Oceanography, University of California, San Diego, California

Abstract

Desert fishes, such as members of the genus *Cyprinodon,* and estuarine fishes, such as *Gillichthys mirabilis,* experience and tolerate wider ranges of cell temperature than almost any other vertebrate. These fishes are also remarkably euryhaline.

Two types of enzymic adaptation strategies can be envisioned to permit tolerance of wide ranges of temperature, salinity, and other environmental factors that can perturb enzymic function and structure. First, during evolutionary adaptation to a particular environment, selection may favor the development of enzymes (and respiratory and structural proteins) that are capable of functioning well over the full spectrum of habitat conditions the organism is apt to encounter. Such proteins can be termed "eurytolerant" proteins.

Second, an organism may broaden its environmental tolerance range through the acquisition of additional protein variants, isozymes or allozymes, which supplement the form of the protein already encoded in the genome. This "multiple variant" adaptation strategy therefore involves sets of two or more variants of a particular type of protein, which, when working in conjunction with each other, perform the functions demanded of the protein over the entire environmental range. Each individual protein variant is, however, "stenotolerant": it is not able to function satisfactorily under all environmental conditions.

We have followed two approaches to determine the relative importance of these two adaptational strategies in eurythermal desert and estuarine fishes. Using standard electrophoretic procedures, we examined 14 species of teleosts from widely different thermal regimes to determine whether detectable levels of genetic (protein) polymorphism are positively correlated with the range of cell temperatures the organisms experience. We found no relationship between the range of environmental temperature variability and protein polymorphism. The two eurythermal and euryhaline desert and estuarine fishes, *C. macularius* and *G. mirabilis,* respectively, were found to be low-to-average in their levels of enzyme polymorphism. Tropical fishes and a deep-sea species were highest in polymorphisms. These findings suggest that differences in environmental thermal stability are not sufficient to promote differences in average levels of genetic (enzyme) polymorphism, at least among different species of fishes.

The second mode of analysis used in our work involves comparative studies of enzyme kinetic properties, which are both highly critical for enzymic function and extremely temperature sensitive. Enzyme–substrate affinity, as measured by the apparent Michaelis constant (K_m) of substrate, is an especially important parameter in both regards. Our results, and other observations in the literature, are consistent with the hypothesis that enzyme–substrate binding ability is an extremely sensitive trait in enzyme evolution. The effect of assay temperature on enzyme–substrate affinity differs for each homologue of a given enzyme. Stenothermal fishes possess enzymes that bind substrate well only over a narrow range of temperatures. In contrast, pyruvate kinases of *G. mirabilis* and *C. macularius* bind substrate in a relatively temperature independent manner over a very wide range of temperatures and, in addition, are capable of binding substrate very well at temperatures above 25°C, a trait that is not characteristic of pyruvate kinases of less warm-adapted fishes. In fact, the pyruvate kinase of *C. macularius* increases its substrate binding ability as the temperature of assay is raised from 10 to 40°C, a unique trait among all ectothermic pyruvate kinases studied to date. The enzymes of these highly eurythermal fishes therefore display the required catalytic and regulatory properties to ensure that enzyme function is maintained in the face of extremely high and highly variable cell temperatures.

THERMAL STRESS IN DESERT AND ESTUARINE HABITATS

The severity of the stress imposed upon an organism by a change in cell temperature is proportional to the product of two terms: (1) the magnitude of the temperature change, and (2) the rate at which this change occurs. If the rate of temperature change is extremely slow relative to the organism's generation time, the processes of genetic mutation and selection may lead to the acquisition of the necessary new genetic information to code for the biochemical adaptations required for success under the new thermal regime. Indeed, the success with which ectothermic (poikilothermic) organisms have colonized such "adverse" environments as the polar seas and desert streams indicates that, given sufficient numbers of generations, ectotherms can develop extraordinary tolerances for both extremes of absolute temperature and large-scale diurnal and seasonal fluctuations in temperature.

Temperature changes of these latter two time courses, i.e., changes which are rapid relative to the organism's generation time, appear to pose the most severe threats to organisms. Since for multicellular organisms no new genetic information can be accumulated during the periods of these thermal changes, successful adaptation can only be effected if (1) the organism's phenotype is inherently temperature tolerant, and/or (2) the genetic repertoire of the organism contains sufficient information to enable the organism to produce new types of biomolecules, e.g., enzymes and lipids, to compensate for the environmental stress. In this paper we shall discuss what appear to be two of the potentially most important mechanisms for effecting phenotypic modifications to cope with large and/or rapid changes in cell temperature.

In elucidating biological principles, a critical first step in one's analysis often involves the selection of an advantageous experimental organism. For study of the biochemical mechanisms of eurythermality, there seem few more suitable study organisms than desert fishes of the genus *Cyprinodon* and the estuarine species, *Gillichthys mirabilis* (the longjaw mudsucker). The desert pupfishes and *Gillichthys* probably experience the greatest ranges of cell temperatures encountered by at least any vertebrate species. In the shallow pools along the shoreline of the Salton Sea, daily temperature variations of 15 to 20°C in summer and winter are common (Barlow, 1958). Temperature fluctuations in shallow desert streams, which often are not appreciably deeper than the pupfish are "tall," may be even greater (Brown and Feldmeth, 1971). The estuarine habitat of *Gillichthys* is somewhat less thermally variable, but again daily temperature changes of 10°C are common, and seasonal variation in temperature may exceed 25°C. Both the desert pupfish and *Gillichthys* likely experience ranges of cell temperature that are very close to the observed ranges of ambient water temperature (Barlow, 1958; Somero, personal observations); avenues of behavioral thermal regulation are often closed for these fishes when they are trapped in shallow pools during low tides or during summer periods of low water level in the desert, conditions under which water temperatures may exceed 40°C.

POTENTIAL ENZYMIC MECHANISMS OF EURYTHERMALITY

The biochemical systems that are instrumental in establishing thermal tolerance limits and thermal optima for organisms are clearly multifarious (Hochachka and Somero, 1973). Two classes of biomolecules seem of particular importance in these regards. Membrane lipids display highly temperature sensitive physical state changes; as a consequence, the membrane systems of the cell are likely both a major locus of thermal "injury" (Bowler et al., 1973; Tansey and Brock, 1972) and an important site of biochemical adaptation (e.g., see Sinensky, 1974).

The other class of biomolecule that plays a crucial role in thermal relationships is the diverse array of enzymic, respiratory, and structural proteins of the cells and body fluids. Since virtually every reaction of metabolism is enzyme-catalyzed, temperature effects on enzyme structure and function are an extremely important determinant of an organism's abilities to withstand any given thermal regime. We shall consider potential mechanisms for establishing tolerance of both (1) extremes of absolute temperature, and (2) large and rapid fluctuations in temperature.

The types of enzymic adaptations that could facilitate tolerance of wide ranges of temperature seem to fall into two categories, as illustrated in Figure 1. To maintain such critical traits as (1) correct substrate binding ability, (2) proper regulatory sensitivities, and (3) required higher orders of structure, in the face of temperature (or other environmental) changes that tend to perturb these traits, selection may lead to the evolution of "eurytolerant" ("eurythermal," "euryhaline," "eurybaric") proteins which are capable of performing all necessary functions over the entire environmental temperature (salt, pressure) range experienced by the organism.

As an alternative to the "eurytolerant protein" strategy, adaptation to a variable environment may involve the acquisition of additional variants on each protein "theme." In the "multiple variant" approach to eurythermality, the organism maintains a set of enzymes within each cell which function cooperatively, as shown in Figure 1. Each individual enzyme variant is "stenothermal"; i.e., it can function well only over part of the full environmental temperature range. However, the joint functioning of all these "stenothermal" variants provides the organism with the same enzymic potential that would be provided by a single "eurythermal" enzyme.

Although on functional grounds both of these adaptational strategies seem of equal merit, it seems probable that the eurytolerant protein strategy offers a more satisfactory solution to coping with variable environments. In terms of genetic considerations, e.g., if two or more

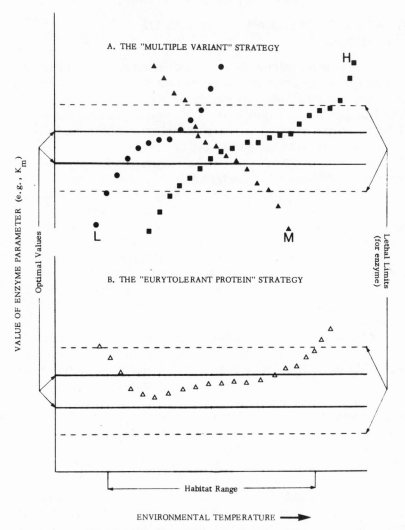

FIGURE 1 Alternative strategies for maintaining enzymic function over a broad range of an environmental parameter such as temperature, which tends to perturb enzymic function and structure. In the multiple variant strategy (A), multiple allozymes and/or isozymes are necessary to maintain the enzymic parameter in question, e.g., the apparent Michaelis constant (K_m) of substrate, within an acceptable range over the entire environmental spectrum experienced by the organism (and enzyme). A low-range (L) variant works optimally at low temperatures; midrange (M) and high-range (H) variants function optimally over the mid- and high-temperature regions. In the eurytolerant protein strategy of adaptation (B), a single protein form is capable of maintaining its functional characteristics within a tolerable range over the entire span of temperatures faced by the organism (enzyme).

loci are needed to sustain a given metabolic function, the genome of the organism will grow more complex. If, on the other hand, the multiple enzyme forms are allelic variants of a single gene locus, difficulties will arise owing to allele segregation during reproduction. Thus, if survival is dependent on heterozygosity, only half the offspring will survive if a single gene locus is involved in the multiple variant adaptation. If several heterozygous loci are critical for survival, it is obvious that major feats of genetic engineering must be accomplished if completely heterozygous offspring are to be produced.

In spite of the a priori attractiveness of the eurytolerant protein strategy, one must ask whether such an entity as a eurythermal protein is possible. Can a single primary sequence of amino acids be expected to perform satisfactorily as both a catalytic and regulatory agent over, say, a temperature range in excess of 35°C? Based on familiarity with the mammalian and bacterial enzyme literature, one might well expect a negative answer. And, in fact, one senses a generally conservative attitude among biologists concerning the environmental tolerance ranges of proteins. Skepticism about the capabilities of single enzyme forms to function satisfactorily over broad ranges of temperature, hydrostatic pressure, salinity, and other environmental variables that can affect enzyme function may, at least in part, explain what seems to be a common attitude among biologists, that environments which are highly variable in time and space provide the conditions that favor accumulation of genetic variation (Levins, 1968; Johnson, 1973, 1974; also see the references given in Somero and Soulé, 1974).

ENZYME POLYMORPHISM AND ENVIRONMENTAL TEMPERATURE VARIABILITY

The hypothesis that variable environments promote genetic polymorphism has led us to examine the relationship between the range of temperatures experienced by fishes and the levels of protein polymorphisms characteristic of the species. Before discussing this particular approach to the study of the relative importance of eurytolerant strategies and multiple variant strategies, it seems wise first to consider some of the theoretical and practical aspects of electrophoretic analysis of protein variation.

One must remember that protein polymorphisms have two genetic bases. When two separate gene loci code for a particular type of polypeptide chain, the gene products are termed isozymes. Familiar examples of isozymes include the skeletal and heart muscle forms of lactate dehydrogenase and the alpha and beta subunits of adult vertebrate hemoglobins.

When different alleles of a single gene exist, the resulting proteins are called allozymes. In the great majority of studies of protein polymorphisms, allozymes have been studied exclusively. The reason for the greater interest in allozymes derives from the fact that the number of isozyme forms of a given protein is usually small (of the order of 2 to perhaps 10). Furthermore, the number of isozyme forms of proteins are remarkably similar among different species. And, as is obvious, generating a new protein variant by duplicating a gene is apt to be a far more difficult feat than the substitution of one or more amino acids into a protein via mutation(s). However, a multiple variant strategy of adaptation could, at least in theory, involve isozymic systems as well as allozymic differences.

The study of allozymic and isozymic polymorphisms requires careful adherence to a number of experimental design rules. In investigations of the relationships between environmental variability and protein polymorphism one must always be certain that the environmental factor of interest is "felt" by the proteins being examined electrophoretically. Thus, for example, variation in food plant composition may be "felt" only by the enzymes involved in the initial processing of ingested food; enzymes remote from these initial digestive reactions will not be "aware" of the source of their metabolites. Arguments such as these offer strong support for the view that temperature is probably the optimal environmental parameter for use in studies of environment–protein interactions. In ectothermic species such as fishes, changes in temperature are felt rapidly and quantitatively by *all* proteins of the body (albeit not all proteins can be expected to be equally temperature sensitive). If temperature is the environmental parameter of interest, one can therefore electrophoretically examine any convenient set of proteins, knowing that these proteins do exist at the organism–environment interface of interest.

Meaningful comparisons of levels of enzyme polymorphism between populations or species dictate that the same enzymes be studied in all groups to be compared. Different classes of enzymes are known to have characteristically different levels of polymorphism (Johnson, 1974; Selander and Kaufman, 1973).

The most difficult aspect of designing experiments to test for relationships between environment variability and genetic polymorphism derives from the fact that many other factors besides environmental heterogeneity contribute to the level of genetic polymorphism in a species. Any effect of environmental variability may be masked by these other influences. Different groups of organisms may have characteristically different levels of polymorphism; e.g., invertebrates have been claimed to possess higher levels of polymorphism than mammals (Selander and Kaufman, 1973), although Johnson (1974) has questioned this interpretation. Population size may also affect the number of alleles in a population if a majority of the alleles are selectively neutral (Kimura and Ohta, 1971). The relative stability of an environment over geological time periods may also affect levels of polymorphism (Soulé, 1972, 1973). Thus, a clean separation of the effects of environmental variability from the effects of many other factors contributing to genetic polymorphism may be extremely difficult.

Bearing these considerations in mind, we estimated the amounts of protein polymorphism in 14 teleost fishes that inhabit extremely diverse thermal habitats. Interspecific comparisons are, of course, a "coarse-grain" analysis of environment–protein interrelationships. However, since the species to be studied included (1) Antarctic fishes, which live at virtually constant temperatures, (2) tropical fishes, which also live in thermally stable habitats, (3) a deep-sea species, which is also exposed to only a very narrow range of temperatures, and (4) estuarine and desert fishes, which experience extraordinarily wide ranges of temperature, we felt that we would be able to detect a significant level of correlation between temperature variability and genetic polymorphism if variation in environmental temperature is a major factor in determining the level of isozymic and/or allozymic complexity of a group of organisms.

The enzymes we examined electrophoretically are listed in Table 1. References to the precise electrophoretic and staining conditions used in these studies are given in Somero and Soulé, 1974. With the exception of the deep-sea species, *Coryphaenoides acrolepis,* all the enzymes listed were resolved in virtually all species. For *Coryphaenoides,* we were able to detect only the following proteins: lactate dehydrogenase, malate dehydrogenase, esterases, α-glycerophosphate dehydrogenase, phosphohexose isomerase, and general protein bands. Our inability to detect additional enzymes was in part because the livers of the specimens had been removed prior to our acquisition of the fish. In addition, the enzymes of this high-pressure-adapted form may be relatively labile.

TABLE 1 Proteins examined electrophoretically[a]

Dehydrogenases	Other
α-Glycerophosphate dehydrogenase	Acid phosphatase
Ethanol dehydrogenase	Esterases
Glucose-6-phosphate dehydrogenase	Fumarase
Isocitrate dehydrogenase	Glutamate-oxaloacetate transaminase
Lactate dehydrogenase	Leucine aminopeptidase
Malate dehydrogenase	Peptidases
6-Phosphogluconate dehydrogenase	Phosphoglucomutase
Octanol dehydrogenase	Phosphohexose isomerase
Xanthine dehydrogenase	General protein

[a] The electrophoretic conditions used are described in Somero and Soulé, 1974.

The levels of enzyme polymorphism detected in the 14 fishes are listed in Table 2. The species are ranked in order of increasing environmental (cell) temperature range. Our data reveal no correlation between the range of temperatures experienced by the organism and the degree of isozymic or allozymic polymorphism. There is no indication of hig. :er numbers of gene loci, i.e., more isozyme forms, in the more eurythermal fishes. Similarly, the extent of allozyme polymorphism, which can best be measured using the mean individual heterozygosity (Het) statistic (which is less sample-size-dependent than estimates of percentages of loci that are polymorphic), is not correlated with temperature variability. Statistical testing revealed no differences in polymorphism levels among fishes encountering temperature variations in excess of 5°C relative to species experiencing less variation (Somero and Soulé, 1974).

The finding that both *Gillichthys* and the desert pupfish are less polymorphic than the average of the fishes studied (approximately 5.1 percent average heterozygosity) seems particularly significant in view of the fact that these two species are extremely euryhaline as well as very eurythermal. These two species also must cope with a wide range of oxygen tensions in their native habitats. However, the stresses imposed by wide fluctuations in temperature, salt content, ionic composition, and oxygen availability do not appear to have elicited any considerable reliance on a multiple variant strategy of environmental adaptation. These findings thus hint that adaptation to highly variable environments can be achieved by an alternative biochemical strategy, the eurytolerant protein strategy. Indeed, the kinetic data that follow and theoretical arguments phrased on genetic grounds argue very strongly that the eurytolerant protein solution to the problems raised by variable environments is apt to be an optimal mechanism of adaptation.

FUNCTIONAL EVIDENCE FOR EURYTHERMAL PROTEINS

Studies of biochemical adaptation to the environment obviously cannot rely solely on electrophoretic descriptions of proteins, for the true test of whether or not a protein variant is "adaptive" can only come from examining the functional traits of the molecule. It is now certain that a major site of environmental impingement on enzyme function is the set of enzyme-ligand interactions, which includes enzyme-substrate, enzyme–modulator, and enzyme–cofactor pair formations. Both the rate and direction of catalysis are strongly determined by enzyme–substrate and enzyme–modulator interactions. And, since these enzyme–ligand complexes are stabilized in most cases by a small number of "weak" bonds (hydrogen bonds, electrostatic interactions, or hydrophobic interactions), changes in temperature, salt content/composition, or hydrostatic pressure can be expected to exert large-scale influences on metabolism.

Enzyme–substrate interactions have received the greatest amount of attention in terms of environmental effects (Baldwin, 1971; Baldwin and Hochachka, 1970; Hazel, 1972; Hochachka and Lewis, 1970; Moon and Hochachka, 1971, 1972; Somero, 1969a, b; Somero and Hochachka, 1968, 1971). Studies of several enzymes from a diverse set of ectothermic species have shown that enzyme–substrate affinity (1) is extremely similar among interspecific homologues of an enzyme over each species' normal adaptation temperature range, and (2) although gradual increases in enzyme–substrate affinity with decreasing temperatures may serve to reduce in situ temperature coefficients, rapid changes in enzyme–substrate affinity with temperature are generally avoided within a species' normal environmental temperature range (Hochachka and Somero, 1973).

These relationships, as well as evidence that homologues of a given enzyme may differ markedly in their eurythermality, are shown in Figures 2 and 3. The data from studies of acetylcholinesterases (AChE) of brain tissue of different fishes (Baldwin, 1971; Baldwin and Hochachka, 1970) illustrate very clearly the role of enzyme–substrate affinity (as estimated by the apparent Michaelis constant K_m) in the process of temperature adaptation. The K_m versus temperature function for each homologue of AChE is distinct. Most AChE variants exhibit highest affinity for substrate, acetylcholine (ACh), at temperatures near the species' normal habitat temperature. (This statement should not be interpreted to mean that the smaller the K_m value, the better the enzyme, however.) Large and rapid changes in K_m usually occur only at or beyond the normal temperature limits of the organism.

Another important difference among the AChEs studied is the variation in eurythermality among the interspecific homologues. As defined in Figure 1, an enzyme that is able to maintain the value of a parameter at a relatively stable level over a wide range of environmental conditions is a eurytolerant enzyme. By this criterion, the AChE of, e.g., the mullet *(Mugil cephalus)*, a very eurythermal fish (Table 2), is a eurythermal enzyme. And, since Baldwin (1971) has shown that only a single electrophoretic form of AChE accounts for each curve in Figure 2, we can be quite certain that the mullet has adopted the eurytolerant protein strategy, not the multiple variant strategy.

In sharp contrast to the mullet AChE, the homologous enzyme of the stenothermal Antarctic fish *Trematomus borchgrevinki* displays a rapid loss in AChE binding ability over a narrow span of temperatures. The rapid loss in substrate affinity of AChE at temperatures only a few degrees above 0°C may in part account for the low upper lethal tempera-

TABLE 2 Levels of enzyme polymorphism in teleost fishes from different thermal regimes

Species (family)	Collection site (habitat)	Annual[a] thermal range (°C)	Sample size	Loci surveyed	Loci polymorphic[b] (%)	Loci (%)	Het[c] ± S.E. (%)
Trematomus borchgrevinki (Nototheniidae)	McMurdo Sound, Antarctica (subice, pelagic)	1	9	21	4.8	0	0.5 ± 0.5
Trematomus hansoni (Nototheniidae)	McMurdo Sound (benthic)	1	26	26	18.5	11.1	2.5 ± 0.4
Trematomus bernacchii (Nototheniidae)	McMurdo Sound (benthic)	1	30	26	42.3	15.4	3.3 ± 0.6
Dascyllus reticulatus (Pomacentridae)	Manila (tropical reef, shallow)	3	10	29	34.5	—[d]	10.7 ± 1.3
Amphiprion clarkii (Pomacentridae)	Manila (tropical reef, shallow)	3	11	27	22.2	—[d]	9.1 ± 1.1
Halichores sp. (Labridae)	Manila (tropical reef, shallow)	3	10	28	21.4	—[d]	5.7 ± 1.5
Coryphaenoides acrolepis (Macruridae)	San Diego Trough (deep benthic)	5	18	6	66.7	—[d]	11.0 ± 2.8
Abudefduf troschelii (Pomacentridae)	Revillagigedo Islands (tropical, shallow)	7	16	20	35.0	15.0	5.0 ± 1.1
Leuresthes tenuis (Atherinidae)	San Diego (inshore, pelagic)	10	20	33	18.2	15.2	3.6 ± 0.7
Gibbonsia metzi (Clinidae)	San Simeon, California (benthic, intertidal)	10	28	28	28.6	17.9	4.3 ± 0.8
Bathygobius ramosus (Gobiidae)	Revillagigedo Islands (tropical, shallow)	12	16	23	4.3	4.3	0.5 ± 0.4
Mugil cephalus (Mugilidae)	Mission Bay, San Diego (inshore, pelagic, tropical and subtropical)	15	20	30	36.7	20.0	7.1 ± 1.1

Gillichthys mirabilis (Gobiidae)	San Diego (benthic, estuarine)	20	30	29	30.0	20.0	4.6 ± 1.3
Cyprinodon macularius[e] (Cyprinodontidae)	Salton Sea (drainage canals, pools)	35	18	21	23.8	23.8	3.3 ± 0.8

[a] The annual thermal range is an approximation of the variation in water (body) temperature a member of the sampled population would experience in its native habitat. These estimates are based on published water temperature data and personal observations.
[b] When commonest allele has a frequency of less than 0.95.
[c] Mean individual heterozygosity, ± the standard error of the estimate.
[d] Samples too small to permit calculation of parameter.
[e] All data, other than those for *C. macularius*, are from Somero and Soulé (1974).

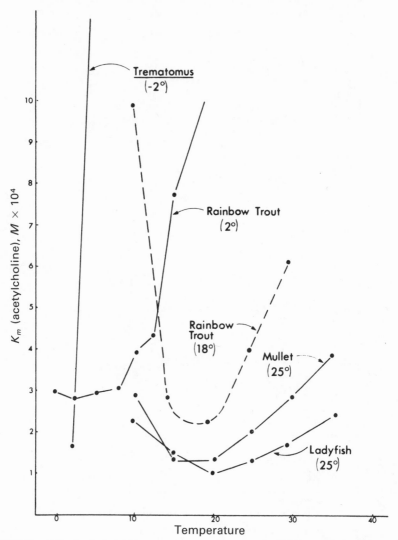

FIGURE 2 Effect of assay temperature on the apparent Michaelis constant (K_m) of acetylcholine for acetylcholinesterases from brain tissue of several species of fishes. Experimental methods are described in Baldwin and Hochachka, 1970, and Baldwin, 1971. (From Hochachka and Somero, 1973, with permission of the authors.)

ture of this species, approximately 7°C (Somero and DeVries, 1967). Organismal stenothermality may thus be due to the stenothermality of one or more key enzymes such as AChE.

A final observation of some importance is that the rainbow trout *(Salmo gairdnerii)* seems to differ from the other fishes studied by favoring the eurytolerant protein strategy both kinetically and elec-

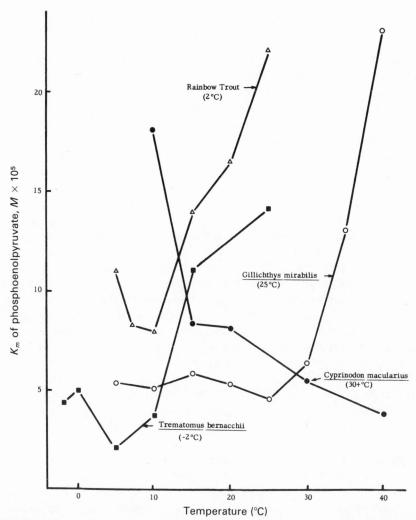

FIGURE 3 Effect of assay temperature on the apparent Michaelis constant (K_m) of phosphoenolpyruvate (PEP) for pyruvate kinases (PKs) from different fishes. Enzyme assay techniques are described in Somero and Hochachka, 1968. Data for the PKs of rainbow trout, *Trematomus bernacchii,* and *Gillichthys mirabilis* are from Hochachka and Somero, 1973. The K_m values for PK of *Cyprinodon macularius* were determined using 18,000 × g supernatants of skeletal muscle homogenates (prepared with 5 volumes of 0.05 M tris/HCl buffer, pH 7.5). K_m values were determined by computer from plots of velocity/substrate concentration versus substrate concentration.

trophoretically. In the rainbow trout two distinct AChE isozymes can be detected. Cold-acclimated trout possess an AChE that binds substrate relatively well at low temperatures but loses its high substrate affinity at summer temperatures. Warm-acclimated trout contain a different enzyme that binds substrate better at high, summer temperatures than at low, winter temperatures. Trout acclimated to intermediate temperatures (10 to 12°C) contain both isozymes (Baldwin and Hochachka, 1970). Rainbow trout therefore seem to opt for a multiple variant (multiple *iso*zyme) strategy of adaptation, in contrast to other fishes. The basis for this difference may derive from the atypical genetics of the rainbow trout, since these organisms are thought to be recent tetraploids (Ohno, 1970). The presence of a twofold dose of enzyme-encoding gene loci has seemingly provided the trout with the raw material for elaborating season-specific isozyme forms (Somero, 1975). Fishes lacking this genetic raw material appear to have the multiple isozyme avenue of adaptation closed to them, at least in terms of being able to generate isozymes for a large percentage of the cellular proteins.

Another enzyme that has been examined in a number of ectothermic species is skeletal muscle pyruvate kinase (PK) (Fig. 3). The interspecific differences among PK homologues are similar to the differences noted for AChE homologues. Substrate (phosphoenolpyruvate) affinity is normally very similar among the interspecific homologues of the enzyme when comparisons are made at the physiological temperatures of the organisms. Large and rapid changes in enzyme–substrate affinity within the species' normal temperature range are uncommon. Again, the rainbow trout appears to possess two, season-specific isozymes of PK (Somero and Hochachka, 1971).

The muscle PKs of *Gillichthys* and the pupfish are interesting in several respects. Both PKs merit the title eurythermal enzyme since their PEP binding abilities are relatively temperature insensitive over an extremely wide range of temperatures. The pupfish enzyme is especially striking, relative to the other fish PKs, in that it is unique in displaying an increased affinity for PEP as temperature is raised from 10 to 40°C. There can be little doubt that the pupfish PK is both a warm-adapted and a eurythermal enzyme. It would be most interesting to determine whether other enzymes of this heat-tolerant fish attain their maximal substrate binding ability at such high temperatures. It should also be noted, however, that the pupfish enzyme appears to be a relatively poor enzyme at temperatures at the low end of its environmental spectrum (below approximately 15°C). At these low temperatures the enzyme begins to lose its PEP binding ability quite rapidly, and this decrease as thermal energy is also decreasing can lead to enormous temperature coefficients at physiological substrate concentrations (Somero and Hochachka, 1971).

In the case of the *Gillichthys* and pupfish PKs, we have not determined the number of isozyme forms present in muscle tissue, so our conclusion that the PKs of these species are single, eurythermal enzymes is tentative. However, we have fully purified skeletal muscle PKs from several other fishes (Low and Somero, unpublished), and in each case a single isozyme is present. This evidence, plus the finding that neither *Gillichthys* nor the pupfish contains greater than average numbers of gene loci for other proteins (Table 2), serves, we feel, as a strong base to make an inductive prediction that the muscle PKs of both species exist as single molecular forms.

DISCUSSION

On the basis of our results we conclude that adaptation to variable environments can be achieved, for at least some enzymes in lower vertebrates, without increasing levels of isozymic or allozymic polymorphism. Kinetic studies suggest that a single enzyme, in terms of primary sequence, can function satisfactorily over a very broad range of temperatures. Eurythermal organisms appear to have especially eurythermal enzymes, relative to the homologous enzymes of less thermally tolerant species.

There are several lines of reasoning which lead us to conclude that the eurytolerant protein strategy of adaptation to variable environments is an optimal strategy. Arguments based on genetic grounds, e.g., genome simplicity and segregational mechanics, seem to strongly favor a single-protein approach to environmental adaptation. In addition, one can phrase objections to a multiple variant strategy in terms of the cellular energy economy and the architectural framework of the cell. If an organism is synthesizing two forms of an enzyme at all times, as is generally true in the case of allozyme systems, except in rare cases such as the Lyon effect, it may find that its cells contain essentially "useless" enzyme molecules at one or the other extreme of the habitat range. For example, a cold-functioning allozyme may contribute nothing but clutter to the cell during the summer, and its synthesis will serve as a rewardless drain on the cell's energy budget. Perhaps more critically, the solvent capacity of the cell (Atkinson, 1969) and the sites for binding enzymes to membranes and other organelles are limited in number. Large numbers of nonfunctioning or poorly functioning allozymes would thus be at a disadvantage on these two grounds as well.

Although the eurytolerant protein strategy of adaptation might appear to be the ideal mechanism of adaptation to environments which vary in time and/or space, it is clear that not all organisms have adopted or achieved this strategy for all their gene loci. There are a great many well-established cases in which multiple allozyme or multiple isozyme systems appear to play a role in environmental adaptation. However, it should be stressed that these polymorphism–environment correlations usually are genetic observations, not studies in which functional differences between allozymes or isozymes have been demonstrated.

Nonetheless, if the eurytolerant protein strategy seems so good a priori, we must face the question as to why it is not the universal choice among enzyme systems.

In the case of allozyme polymorphisms, there are at least two reasons why an organism may use a multiple variant strategy rather than a eurytolerant protein strategy. First, and most simply, there may be enzymes, respiratory proteins, and structural proteins for which structural limitations prevent acquisition of broad environmental tolerances. Thus, whereas the enzymes we have studied kinetically appear capable of acquiring impressive degrees of eurythermality over evolutionary time spans, there may well be other classes of proteins that, for reasons which are not now clear, are able to function only over relatively narrow ranges of temperature. To date, we are unaware of any data bearing on this possibility.

A second reason why multiple variants may be used to broaden an organism's environmental tolerance range stems from a consideration of the influence of time on the adaptation process. Extensions of a protein's environmental tolerance range will almost inevitably lead to at least transitory increases in levels of enzyme polymorphisms, since the evolutionary modification of a protein to extend its tolerance ranges will involve amino acid substitutions at different rates and/or at different locations in the primary strctures of the two allelic forms of the protein. In an admittedly teleological sense, we can propose that enzymes which are evolving "toward" broader environmental tolerance ranges will of necessity display increased amounts of polymorphism. Perhaps some of the cases in which polymorphisms are most prevalent reflect systems that are in the process of gaining eurytolerance to different environmental factors.

The other type of multiple variant strategy we have witnessed involves increased isozymic, rather than allozymic, complexity. The use of multiple isozymes rather than multiple allozymes has distinct advantages, of course. Since different isozymes can be selectively synthesized, the organism need not always maintain non- or poorly functioning proteins within its cells, as we argued was the case in a multiple allozyme system. Thus, energy waste and "packaging" problems are less important than in an allozyme system. The element of time is very important in gaining an appreciation of the likely roles of multiple isozyme systems. In eukaryotes, the activation of a gene and the synthesis of a new type of enzyme usually involves time periods of at least hours. In the rainbow trout, the appearance of new isozymes occurs only after 2 to 3 weeks of acclimation (Baldwin, personal communication). Ontogenetic changes in isozymes and respiratory proteins may occur over even longer time courses. These observations lead us to suggest that multiple isozyme adaptation strategies will be found to characterize certain processes that occur relatively slowly, e.g., seasonal adaptations. It is much less likely that changes in isozyme populations would, or even could, occur in response to diurnal changes in the environment.

In the case of the isozyme adaptations found in the rainbow trout, which to date appears to be the only species in which clearly defined isozyme changes related to temperature occur, one is tempted to ask "why" this species has opted for this particular mode of adaptation. Perhaps the answer to this riddle will come from study of plant systems, where, as in the case of the rainbow trout and other Salmonid fishes, polyploidy is often correlated with an ability to tolerate wide ranges of environmental conditions. It seems likely, in fact, that in situations where increases in ploidy can occur with relative ease, e.g., in plants, where sex determination mechanisms do not act as a prohibition against chromosomal duplication, the multiple variant strategy may be of considerable importance. The nature of the genetic system with respect to sex determination mechanisms may be a vital underlying determinant of the adaptational strategy used by an organism.

In conclusion, let us consider some of the questions which these studies have raised concerning adaptations of the desert pupfish. The data presented in this paper represent but a very shallow scratch on the surface of desert and estuarine fish biochemistry. Thus, we know virtually nothing about the other 99.9 percent of the pupfish's enzymes, and we have no insights to date as to the structural bases of the thermophilic and eurytolerant kinetic properties of the pupfish pyruvate kinase. In addition to seeking answers to these questions through further study of *C. macularius*, curiosity leads one to inquire about possible evolutionary differences between the pupfish, such as *C. macularius*, which live in thermally variable habitats, and the closely related pupfish, such as *C. diabolis*, which live at virtually constant temperatures. Evolution under stable thermal conditions appears to have had only a minimal effect on the eurythermality of *C. diabolis* (Brown and Feldmeth, 1971), and Turner (1972, 1973a, b, 1974) discovered remarkable electrophoretic similarities among enzymes of several pupfish species. Perhaps the ancestral pupfish possessed a well-developed set of eurythermal proteins, and this set of proteins has been retained in all present-day pupfish species and populations.

A final question concerning enzymic adaptation in the desert pupfish would appear to cut to the core of the rationale for doing comparative biochemistry. Enzymes of the pupfish, birds, and mammals must function at approximately the same temperatures. Has selection therefore led to a convergent pattern of evolution, say in the amino acid compositions of the homologous enzymes of these different groups? How do the higher orders of structure, and the bonds stabilizing these structures, compare in these three groups? Questions such as these clearly show that the comparative biochemist has important fundamental contributions to make to the study of life. By studying the variations among homologues of a protein and the common "solutions" used by proteins to cope with particular problems, one can more clearly discern what traits are absolutely essential for molecular structure, how these traits are maintained through adjustment of the amino acid composition of a protein, and, as a result of such analyses, discover the mechanisms of evolution that operate at a basic, molecular level.

Acknowledgments

This research was supported by NSF Grant GB-31106. I gratefully acknowledge stimulating discussions of these ideas with Bonnie Jean Davis, Michael Soulé, and Christopher Wills.

LITERATURE CITED

Atkinson, D. E. 1969. Limitation of metabolite concentrations and the conservation of solvent capacity in the living cell. In B. L. Horecker and E. B. Stadtman (eds.), Current topics in cellular regulation, Vol. 1, pp. 29–43. Academic Press, New York.

Baldwin, J. 1971. Adaptation of enzymes to temperature: acetylcholinesterases in the central nervous sytem of fishes. Comp. Biochem. Physiol. 40:181–187.

————, and P. W. Hochachka. 1970. Functional significance of isoenzymes in thermal acclimation: acetylcholinesterase from trout brain. Biochem. J. 116:883–887.

Barlow, G. 1958. Daily movements of desert pupfish, *Cyprinodon macularius,* in shore pools of the Salton Sea, California. Ecology 39:580–587.

Bowler, K., C. J. Duncan, R. T. Gladwell, and T. F. Davison. 1973. Cellular heat injury. Comp. Biochem. Physiol. 45A:441–450.

Brown, J. H., and C. R. Feldmeth. 1971. Evolution in constant and fluctuating environments: thermal tolerances of desert pupfish *(Cyprinodon).* Evolution 25:390–398.

Hazel, J. R. 1972. The effect of temperature acclimation upon succinic dehydrogenase activity from the epaxial muscle of the common goldfish (*Carassius auratus* L.): lipid reactivation of the soluble enzyme. Comp. Biochem. Physiol. 43B:863–882.

Hochachka, P. W., and J. K. Lewis. 1970. Enzyme variants in thermal acclimation: trout liver citrate synthases. J. Bol. Chem. 245:6567–6573.

————, and G. N. Somero. 1973. Strategies of biochemical adaptation. W. B. Saunders, Philadelphia. 358 p.

Johnson, G. B. 1973. Enzyme polymorphism and biosystematics: the hypothesis of selective neutrality. Ann. Rev. Ecol. System. 4:93–116.

————. 1974. Enzyme polymorphism and metabolism. Science 184:28–37.

Kimura, M., and T. Ohta. 1971. Protein polymorphism as a phase of molecular evolution. Nature 229:467–469.

Levins, R. 1968. Evolution in changing environments. Princeton University Press, Princeton, N.J.

Moon, T. W., and P. W. Hochachka. 1971. Temperature and enzyme activity in poikilotherms: isocitrate dehydrogenases in rainbow-trout liver. Biochem. J. 123:695–705.

————, and P. W. Hochachka. 1972. Temperature and the kinetic analysis of trout isocitrate dehydrogenase. Comp. Biochem. Physiol. 42B:725–730.

Ohno, S. 1970. Evolution by gene duplication. Springer-Verlag, Berlin.

Selander, R. K., and D. W. Kaufman. 1973. Genic variability and strategies of adaptation in animals. Proc. Nat. Acad. Sci. (U.S.) 70:1875–1877.

Sinensky, M. 1974. Homeoviscous adaptation—A homeostatic process that regulates the viscosity of membrane lipids in *Escherichia coli.* Proc. Nat. Acad. Sci. (U.S.) 71:522–525.

Somero, G. N. 1969a. Pyruvate kinase variants of the Alaskan king crab: evidence for a temperature-dependent interconversion between two forms having distinct and adaptive kinetic properties. Biochem. J. 114:237–241.

————. 1969b. Enzymic mechanisms of temperature compensation: immediate and evolutionary effects of temperature on enzymes of aquatic poikilotherms. Amer. Nat. 103:517–530.

————. 1975. The role of isozymes and allozymes in adaptation to varying temperatures. In C. L. Markert (ed.), Proceedings of the Third International Conference on Isozymes, Academic Press, New York. In press.

————, and A. L. DeVries. 1967. Thermal tolerance of some Antarctic fishes. Science 156:257–258.

————, and P. W. Hochachka. 1968. The effect of temperature on catalytic and regulatory functions of pyruvate kinases of the rainbow trout and the Antarctic fish *Trematomus bernacchii.* Biochem. J. 110:395–400.

————, and P. W. Hochachka. 1971. Biochemical adaptation to the environment. Amer. Zool. 11:159–167.

————, and M. Soulé. 1974. Genetic variation in marine fishes as a test of the niche-variation hypothesis. *Nature* 249:670–672.

Soulé, M. 1972. Phenetics of natural populations. III. Variation in insular populations of a lizard. Amer. Nat. 106:429–446.

————. 1973. The epistasis cycle: a theory of marginal populations. Ann. Rev. Ecol. System. 4:93–116.

Tansey, M. R., and T. D. Brock. 1972. The upper temperature limit for eukaryotic organisms. *Proc. Nat. Acad. Sci.* (U.S.) 69:2426–2428.

Turner, B. J. 1972. Genetic divergence and variation of Death Valley pupfish populations. Ph.D. Dissertation. University of California, Los Angeles, Calif.

————. 1973a. Genetic variation of mitochondrial aspartate aminotransferase in the teleost *Cyprinodon nevadensis.* Comp. Biochem. Physiol. 44B:89–92.

————. 1973b. Genetic divergence of Death Valley pupfish populations: species-specific esterases. Comp. Biochem. Physiol. 46B:53–70.

————. 1974. Genetic divergence of Death Valley pupfish species: biochemical versus morphological evidence. Evolution 28:281–294.

THYROID AND ADRENAL FUNCTION IN DESERT ANIMALS: A VIEW ON METABOLIC ADAPTATION

Mohamed K. Yousef

Department of Biological Sciences, University of Nevada, Las Vegas, Nevada, and Laboratory of Environmental Patho-Physiology, Desert Research Institute, University of Nevada, Boulder City, Nevada

Abstract

Processes of mammalian adaptation are presented in terms of control system terminology, i.e., sensor, controller, and controlled variables. Most available data deal primarily with elements involved in the controlled variables; little attention has been given to the controller and its elements. For example, although it is well documented that metabolic rate of desert rodents is lower than predicted from the equation describing metabolic rate and body weight, the controller responsible is poorly understood. One element of the controller, i.e., the endocrine system, is discussed as it relates to metabolic adaptations in rodents occupying desert, woodland, and montane forest habitats.

Available data on thyroid and adrenal cortical secretory activities offer, at least in part, a reasonable mechanism to explain the low metabolic rate of desert rodents. There seem to be significant differences associated with ecologic distribution in the secretory activity of the thyroid and adrenal cortex glands of various rodents. These hormonal differences appear to coincide reasonably with the different metabolic rates exhibited by various species. For better understanding of the role of heat production, heat loss, and the balance of water and electrolytes in desert mammals, it is necessary to explore the integrative role of the pituitary, thyroid, parathyroid, and adrenal glands in these animals.

INTRODUCTION

Adaptations of animals to desert environment involve the interaction of the nervous, endocrine, neurohumoral, and motor systems, which synergestically bring about many adjustments in balance of body fluids and in the thermoregulatory, cardiovascular, and respiratory systems. With the present knowledge of physiological and behavioral adaptations of desert animals, it seems reasonable to attempt to phrase the various processes involved in terms of control-system terminology, i.e., sensor, controller, and controlled variables. This suggested system may help to stimulate further experimental work directed to achieve a better understanding of the various elements of the system and its operation as a whole.

The sensor subsystem includes skin receptors and internal temperature (i.e., hypothalamic, core, or blood). Elements involved in the other two subsystems are shown in Figure 1. Major effort has been devoted to gaining an understanding of the right side of this model. Physiological and behavioral responses of many animals to the desert have been identified and quantified; several comprehensive reviews and books are available (Schmidt-Nielson, 1964; Dill et al., 1964; Brown, 1968; Hoff and Riedesel, 1969; Yousef et al., 1972; Maloiy, 1972; Robertshaw, 1974; and Folk, 1974). However, our understanding of the left side of the model is incomplete.

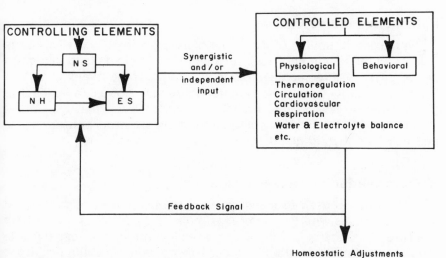

Homeostatic Adjustments

FIGURE 1 Block diagram representing the elements constituting the controlling and controlled subsystems of a feedback control model proposed for the various processes of adaptation. (NS, nervous system; NH, neurohumoral system; ES, endocrine system.)

and their interrelation in the homeostatic adjustments of animals to desert environments. Our purpose is to focus attention on one element of the controller subsystem, the endocrine system, by collecting and integrating available data for hormonal activity in some desert animals, with specific emphasis on rodents. Available data at this time relate primarily to thyroid function and, to a lesser extent, adrenal cortical activity. Therefore, it is appropriate first to review briefly methodology and current evidence on adjustments of the thyroid and adrenal cortex during acclimation and/or acclimatization to heat and cold, and, second, to relate these adjustments to their respective roles in thermoregulation.

METHODS FOR THYROID ACTIVITY

The most frequently employed techniques include those utilizing radioisotopes of iodine: ^{131}I, ^{125}I, and ^{132}I (Burkle and Lund, 1963; Wilson et al., 1971). Selection of the isotope should be guided by instrumentation, radiation exposure, availability, and economy. The physical characteristics of the three isotopes are described in textbooks on radiochemistry. The principles and details of many procedures utilizing radioactive iodine are described by Burkle and Lund (1963) and Luick (1970). A brief introduction to the commonly used methods is given next.

Radioactive Iodine Uptake

The maximum uptake of ^{131}I by the thyroid gland is indicative of the fraction of the iodide pool accumulated by the thyroid gland in a given period of time after administration (usually intraperitonially, i.p.). A hyperactive gland accumulates more iodide than normal; a hypoactive gland takes up less iodide. It should be emphasized that the uptake method reflects the fixation phase and not the secretory and/or release phases of the thyroid gland.

Radioactive Iodine Release Rate

The method is used to measure the rate at which thyroidal labeled iodine decreases with time after maximum uptake. This is not a direct measure of the thyroidal rate of labeled iodine ouput, because this rate is determined partly by the reentry of labeled iodide resulting from the catabolism of the thyroid hormones in tissues. If the reentry rate is blocked, the release rate is likely to be a good index of thyroid activity, at least under normal conditions.

Thyroxine Secretion Rate (TSR)

An estimate of daily TSR can be obtained by two different procedures:

Substitution Method. This method is based on the classical concept that the thyroid and pituitary glands constitute a feedback mechanism in which blood level of the thyroid hormones (thyroxine, T_4, and triiodothyronine, T_3) is the regulator of TSH (thyrotropin stimulating hormone) secretion by the anterior pituitary. The development of this method and the factors affecting it were fully reviewed by Turner (1969). Briefly, the method begins by i.p. injection of a small dose of carrier-free sodium ^{131}I in saline solution. Radioactivity of the thyroid region of the unanesthetized animal is measured. After maximal thyroidal uptake of ^{131}I is reached, exogenous T_4 daily injections begin with a small dose, which is increased daily until a dosage is reached that can block the release of thyroidal ^{131}I (95 to 100 percent of previous daily count). The minimal dose of T_4 that inhibits the release of ^{131}I is equivalent to the daily TSR. A goitrogen, i.e., methimazole (tapazole), must be injected daily throughout the procedure to prevent recycling of ^{131}I from metabolized T_4. A typical example of measuring TSR in two species of desert rodents is shown in Figure 2. Uncertainty of the reliability of this method under certain conditions has been discussed (Heroux and Brauer, 1965).

Degradation rate method. This technique measures the rate of T_4 degradation by the tissues. Assuming a state of homeostatic equilibrium, this rate equals the rate of secretion. A discussion of the application of this method to large and small mammals has been reported (Yousef and Johnson, 1967c, 1968; Yousef and Luick, 1971).

Thyroid Hormones in Blood

After the release of T_4 and T_3 into the blood, both hormones are bound by plasma proteins. In practice, plasma protein bound iodine (PBI) is assumed to be exclusively T_4. Therefore, plasma PBI when multiplied by 1.54 gives the equivalent amount of plasma T_4 since iodine comprises 65 percent of the weight of T_4. The acid distillation method based on the work of Moran (1952) is commonly used. More sensitive and accurate methods based on the principle of the competitive protein binding assay have been developed recently (Sparagana et al., 1969; Ekins et al., 1969; Murphy et al., 1966). Proceedings of a recent workshop dealing with theoretical and practical knowledge of competitive protein binding assays have been published (Odell and Daughaday, 1972). Radioimmunoassay techniques offer a unique and simple procedure for assay of thyroid hormones.

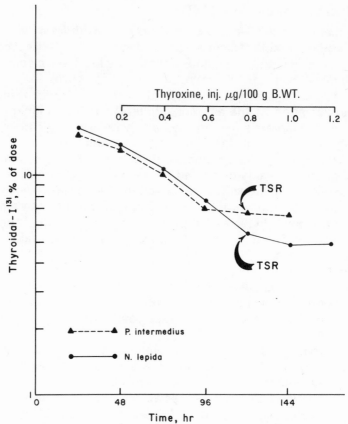

FIGURE 2 Typical example of measuring daily thyroxine secretion rate (TSR) in *Perognathus intermedius* and *Neotoma lepida*. A daily injection of tapazole, 1.0 mg/100 g of body weight, is necessary to block recycling of ^{131}I. Thyroxine doses are injected i.p. in μg/100 g body weight.

METHODS FOR ADRENAL CORTICAL ACTIVITY

The primary secretions of the adrenal cortex include two categories based on physiologic activity: glucocorticoids and mineralocorticoids. The principal glucocorticoids in mammals include corticosterone (compound B) and hydrocortisone (cortisol or compound F). The primary mineralocorticoid is aldosterone. The role of aldosterone in adaptation to heat or cold is incompletely understood; more data are available on glucocorticoids. For this reason the methods presented here on adrenal cortical activity deal primarily with glucocorticoids.

Glucocorticoid Secretion Rate

Isotope dilution technique. This method has not been widely used in small mammals but is used in large mammals (Yousef et al., 1971). Briefly, this procedure involves the determination of concentration of plasma glucocorticoid and disappearance rate of labeled glucocorticoid from the plasma following intravenous injection. Utilizing these measurements and additional calculations, the following equation is used to calculate the daily glucocorticoid secretation rate, GSR:

$$GSR = (K)(PG)(GDS)$$

where PG is the concentration of plasma glucocorticoid, K the fractional disappearance of radioactive injected hormone, and GDS the space in which the labeled hormone presumably is distributed following injection. Details on this method and its principles and terminology have been discussed (Yousef and Johnson, 1967c, 1968; Yousef et al., 1969; Yousef et al., 1971).

Continuous infusion technique. This method has been used only in large animals. Basically, a trace amount of labeled glucocorticoid is infused at a constant rate over a period of hours, which stimulates the entry of endogenous steroid from the gland (Baird et al., 1969). This allows the radioactive and endogenous hormones to attain equilibrium throughout the body. In the steady state the specific activity of the plasma maintains equilibrium, and the amount removed from the circulation per minute is therefore equal to the rate of infusion of radioactivity.

Hormone Levels in the Blood

Several simple fluorometric methods have been developed from the original technique of Silber et al. (1958) and Guilleman et al. (1959). Fluorometric analysis of corticoids has been reviewed recently (Silber, 1966; Chattoraj, 1970; Rubin, 1970). One limitation of this technique is that interfering fluorescence may arise from glassware, solvents, reagents, or from the sample itself (Noujaim and Jeffery, 1970).

Gas–liquid chromatography provides a means for separating and quantifying steroids in one procedure (see Kase, 1970). More recently, competitive protein-binding (CPB) radioassay and radioimmunoassay have received extensive applications for steroid measurements (Powsner and Raeside, 1971; Blahd, 1971; Midgley et al., 1971; Gross et al., 1972).

THYROIDAL AND ADRENAL CORTICAL ASPECTS
OF ADAPTATION TO ENVIRONMENT

Studies on the role of thyroid and adrenal cortical adjustments in adaptation to environment have centered on acclimation and/or acclimatization of man, laboratory, and domestic animals to heat, cold, and altitude. Most of the early studies have been reviewed (Collins and Weiner, 1968; Chaffee and Roberts, 1971; Johnson, 1972, 1974). Adjustments of thyroid and adrenal cortex in response to changes in environmental temperature depend not only on the duration and intensity of exposure to adverse environments, but also on the previous status or acclimation or acclimatization, and to other physiological mechanisms employed by the animal to meet the challenge of the new stressful environment.

To review the literature on thyroidal and adrenal cortical responses to heat and cold is beyond our scope. In general, most data indicate that short-term exposure to heat or cold causes an increase in blood levels of glucocorticoids; long-term, continuous exposure results in a decrease in glucocorticoid levels. Short- and long-term exposure to cold increases thyroid function; however, short-term exposure to heat causes no significant change in thyroid function, whereas long-term exposure decreases it.

What do these hormonal changes mean in terms of adaptation to environment? Thyroid and adrenal cortical hormones have been shown to be calorigenic, even during exposure to heat, in small and large mammals (Evans et al., 1957; Yousef and Johnson, 1966, 1967a; Cassuto and Amit, 1968). Therefore, it is logical that during acclimation to cold both endocrine glands become activated, thus contributing to sufficient increase in heat production to maintain homeothermy. During heat acclimation, when it is beneficial to decrease endogenous heat production, both glands decrease their functional activity, thus helping animals to tolerate heat. Fregley et al. (1963) have demonstrated that rats with a depressed thyroid function were able to tolerate heat better than their controls. Literature in support of this view was reviewed recently in large and small mammals (Hart, 1971; Whittow, 1971; Collins and Weiner, 1968).

Although it is well documented that the physical law of surface-to-volume relationship describes the ratio of metabolic rate (MR) to body weight (the mouse-to-elephant equation) under basal conditions (Brody, 1945; Kleiber, 1961), there are a number of exceptions to this generalization. Acclimation and/or acclimatization to heat and cold results in modification of MR (Hart, 1971; Chaffee and Roberts, 1971; Yousef and Johnson, 1968; Yousef and Dill, 1971a, 1971b). Additionally, a number of small desert mammals have a lower MR than the expected values obtained from the mouse-to-elephant equation (Bradley and Yousef, 1972). This lowered MR is a useful adaptive mechanism since it minimizes the need for food and water, which are at a premium in the desert. Recent work in our laboratory has pointed to a positive correlation between water turnover rate and MR in various desert rodents (Yousef et al., 1974). Also, it is evident that some desert rodents are capable of manipulating their energy budget with availability of food to a degree that these rodents deviate from strict homeothermy under food deprivation (Yousef and Dill, 1971b).

POSSIBLE MECHANISMS OF LOWERED METABOLIC RATE IN SOME DESERT MAMMALS: THYROID AND ADRENAL CORTICAL FUNCTION

Comparative metabolic adaptations were first studied by Scholander et al. (1950). They formulated the hypothesis that, if basal MR of homeothermic animals is determined by an exponential relation to body size, irrespective of climate as shown from Benedict's mouse-to-elephant equation (Benedict, 1938; Kleiber, 1932; Brody and Procter, 1932), and body temperature is maintained constant, then the only factor left to take care of climatic adaptations is insulation. Scholander and his coworkers plotted data on standard MR of arctic and tropical mammals on the mouse-to-elephant curve. Their data indicated that several species of arctic and tropical mammals deviated up to 20 percent, and a few other species deviated even more from the standard curve. Although these deviations are apparent in the data (Scholander et al., 1950, Fig. 1, p. 263), a *tentative* generalization that "basal MR of terrestrial mammals from tropics to arctic is fundamentally determined by a size relation according to the formula cal/day $= 70$ kg$^{0.75}$ and is phylogenetically nonadaptive to external temperature conditions" was suggested. They also suggested that the great deviation of MR in weasels may be indicative of metabolic adaptations, but ruled out such adaptation in the case of the sloth and suggested that the low MR of these animals is tied up with hibernation. This latter mechanism is untenable since many other hibernators do not have a low MR.

Examination of data on small desert mammals, specifically rodents, shows that the MR of many desert species falls below the standard mouse-to-elephant curve; some species have an MR that lines up closely with the standard curve, and others fall well above it (Fig. 3). What then are the mechanisms utilized by these different species for regulation of their MR at the appropriate level? It has been suggested that the low MR of desert rodents probably is related to their high body fat content when kept under laboratory conditions with minimum activ-

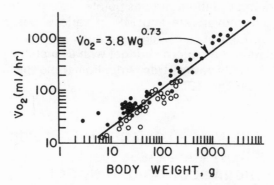

FIGURE 3 Rate of oxygen consumption ($\dot{V}O_2$) for several species of desert and nondesert rodents. The solid line is derived from the equation $\dot{V}O_2$ ml/h = 3.8 Wg$^{0.73}$.

ity and with abundant food (Hayward, 1965). McNab (1969) refuted this hypothesis with his data on two arid-adapted rodents. In our laboratory a study was conducted on the interrelationship of body fat content and MR of 15 species of rodents representing three families and a wide variety of climatic conditions, ranging from low desert to montane forest (Scott et al., 1972). The data are summarized in Figure 4 in which MR and body fat content are related to ecologic distribution. Clearly, low MR is not dependent on body fat content. Furthermore, it seems that ecologic distribution is an important factor affecting MR of rodents.

Since modification of MR during acclimation to environmental temperature, as previously discussed, has been related to changes in the activity of a number of endocrine glands, it is probable that the three different metabolic levels shown in desert rodents are influenced by the endocrine system. Only roles of the thyroid and the adrenal cortical glands in relation to MR of desert rodents have been and are being studied in our laboratory.

Thyroid Function

Hudson and Wang (1969) found that thyroidal ^{125}I release rate is less in *Spermophilus tereticaudus* than *Ammospermophilys leucurus* and suggested that low thyroid function is an adaptation for coping with hot environments. Although this finding correlates with the difference in MR of these two species (Hudson, 1964), the authors did not point out the role of thyroid in low MR of *S. tereticaudus*. They also have shown

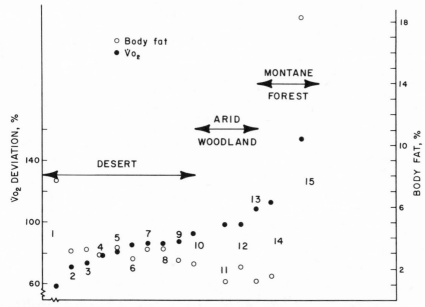

FIGURE 4 Body fat content and rate of oxygen consumption ($\dot{V}O_2$) of 15 species of rodents inhabiting three different ecologic habitats: desert, arid woodland, and montane forest. (Data from Scott et al., 1972.) Each number represents a different species, as shown in the original manuscript.

that during periods of torpidity, thyroidal release rate is decreased. Furthermore, this corresponds with decreased MR during torpidity (Hudson, 1964). Low Thyroid function and low MR have been demonstrated in many other mammals during hibernation (Tashima, 1965; Yousef et al., 1967; Popovic, 1960; Hoffman, 1964).

Thyroid function as measured by several indexes was correlated to MR in 12 species of desert rodents representing three different metabolic levels (Yousef and Johnson, 1972). The data are summarized in Table 1. Data on ^{131}I uptake as an index for thyroid function have been frequently criticized because they can be affected by several factors. The data in Table 1 show that ^{131}I uptake technique is not an adequate index for desert rodents, although in most cases the low ^{131}I uptake coincided with low thyroxine secretion rate (TSR). Using the N/T ratio index (the ratio of ^{131}I counts of the thyroid region, i.e., neck, and of the thigh region), which is used clinically for diagnostic purposes (Burkle and Lunds, 1963), we found no correlation with data on N/T ratio and on ^{131}I uptake or TSR.

To better compare the 12 species studied, TSR values were expressed in micrograms per 100 grams of body weight. The TSR ranged from a low value of 0.63 in *Perognathus formosus* to a high value of 1.35 in

TABLE 1 Daily thyroxine secretion rate, and other related thyroid indices[a]

Species	I^{131} 24-h uptake (%)	N/T ratio	TSR ($\mu g/100\,g$ BW)
Dipodomys merriami	9.5 ± 1.3	4.3 ± 0.2	0.89 ± .24
Dipodomys microps	15.0 ± 0.9	4.6 ± 0.3	0.98 ± .08
Dipodomys deserti	14.7 ± 0.5	5.0 ± 0.2	0.96 ± .32
Perognathus formosus	7.8 ± 0.5	2.6 ± 0.1	0.63 ± .09
Perognathus longimembris	6.4 ± 0.4	3.8 ± 1.0	0.75 ± .16
Perognathus intermedius	11.4 ± 1.7	2.7 ± 0.7	0.80 ± .14
Neotoma lepida	12.6 ± 1.1	6.3 ± 1.0	1.00 ± .09
Peromyscus eremicus	18.8 ± 0.9	2.6 ± 0.2	1.04 ± .10
Peromyscus maniculatus	20.6 ± 2.2	2.7 ± 0.2	1.33 ± .11
Spermophilus tereticaudus	7.5 ± 0.9	5.8 ± 0.7	0.74 ± .07
Spermophilus lateralis	11.9 ± 2.4	4.7 ± 1.0	1.35 ± .10
Ammospermophilus leucurus	9.2 ± 1.0	4.8 ± 0.5	1.07 ± .12

[a] ±, standard error; TSR, thyroxine secretion rate; N/T, neck-to-thigh count ratio; BW, body weight.

Spermophilus lateralis. If one then correlates MR and TSR, as shown in Figure 5, it appears that low TSR coincides with low MR, and vice versa. This suggests that the low MR of desert rodents is at least partially caused by low thyroid function.

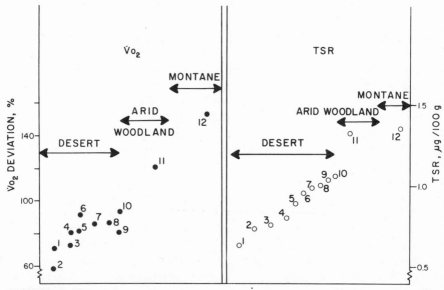

FIGURE 5 Rate of oxygen consumption ($\dot{V}O_2$) and daily thyroxine secretion rate (TSR) of 12 species of desert rodents inhabiting desert, arid woodland, and montane forest areas in the Mohave desert. (Data from Yousef and Johnson, 1972.) Numbers represent the same species in the same order as shown in Table 1.

In cooperation with colleagues at the University of Missouri (I. M. Scott, H. D. Johnson, and M. K. Yousef, unpublished data), plasma thyroxine levels (T_4) were measured in 14 species of desert mammals representing three rodent families and one species of lagomorph. The data summarized in Figure 6 show that species with T_4 levels below 55 ng/ml are those restricted to low desert regions, and the species having T_4 levels above 55 ng/ml either had wide ecological ranges or were found in higher elevations; also it appears that T_4 levels are related to MR. The data on T_4 support the finding on TSR, although for seven species Yousef and Johnson (1972) were not able to see a good correlation between TSR and T_4 levels.

Most T_4 secreted after its release from the gland into the circulation is bound to plasma proteins. The question arises as to whether or not there are differences in the binding of T_4 by plasma proteins in desert rodents. Yousef and Johnson (1972) examined this question and reported that in five species there were three different patterns of T_4 binding, as shown in Figure 7.

1. Albumin and post albumin in *N. lepida* and *D. deserti*.
2. Albumin and alpha 2 in *A. leucurus*.
3. Albumin and interalpha globulins in *S. lateralis* and *S. tereticaudus*.

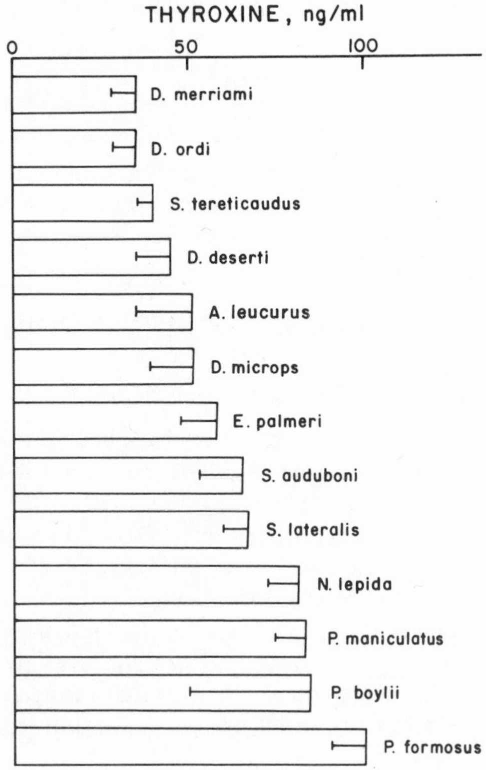

FIGURE 6 Plasma thyroxine concentration of various species of desert animals. (Unpublished data supplied by Scott, Johnson, and Yousef.)

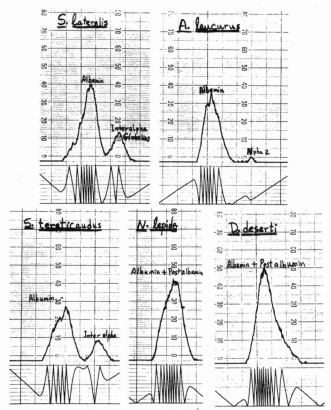

FIGURE 7   Binding sites of thyroxine by plasma proteins in five species of desert rodents. (Data from Yousef and Johnson, 1972.)

The binding patterns of T_4 do not relate to MR nor to ecologic distribution. The significance of this finding is not clear at this time.

Adrenal Cortical Function

Data on the secretory function of the adrenal cortex in desert animals is limited to two studies: one dealt with six species of rodents from the Mohave desert (Vanjonack et al., 1975) and the other with two species of rodents from Egypt (Haggag and El-Husseini, 1974). The adrenal cortex produces two main glucocorticoids: cortisol or hydrocortisone and corticosterone. Which one of these two glucocorticoids is secreted by desert rodents? Plasma glucocorticoids from six species of desert rodents representing three families and different ecologic distribution were determined using the competitive protein binding technique. Plasma corticosterone comprised nearly 100 percent (99.6 ± 0.7 percent) of the total glucocorticoids in all six species (Vanjonack et al., 1975). This finding agrees with other studies on ground squirrels and deer mice (Eleftheriou, 1964; and Huibreytse et al., 1971). Plasma corticosterone levels differed among all species; actual values are shown in Figure 8. Three of the species restricted to low desert had the lowest plasma corticosterone: *Dipodomys deserti, D. merriami,* and *Spermophilus tereticaudus.* Although blood samples were taken between the hours of 10:00 A.M. and 4:00 P.M., it is evident that diurnal species (ground squirrels) had higher plasma corticosterone levels as compared to the nocturnal species (kangaroo rats and desert wood rat). With the exception of one species, *Ammospermophilus leucurus,* it seems that corticosterone levels are related to ecologic distribution: rodents inhabiting low desert areas have lower plasma corticosterone than rodents found in high deserts or montane forest.

What does the low adrenal cortical secretion mean in terms of metabolic adaptations? The answer must be speculative because of the lack of data. It is possible that the low MR in rodents inhabiting low desert areas is related, at least in part, to decreased adrenal cortical secretion since glucocorticoids have been shown to be calorigenic hormones in large and small animals (Evans et al., 1957; Yousef and Johnson, 1967a). This suggestion is supported by the work of Lemaine et al. (1965), who demonstrated that the decrease in plasma levels of glucocorticoids is a specific adaptive response and not due to exhaustion of the adrenal cortex. A decrease in the level of the calorigenic hormones would result in lowering the animal's endogenous heat production, thus enabling it to tolerate heat during summer. Additionally, low corticosterone levels may be associated with water conservation,

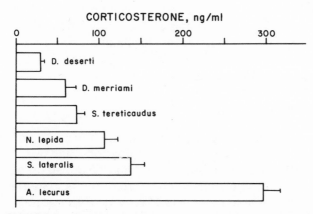

FIGURE 8 Plasma corticosterone concentration of six species of rodents inhabiting the Mohave desert. (Data from Vajonack et al., 1974.)

since it was found that the inhibitory effect of norepinephrine (NE) on the action of antidiuretic hormone (ADH) on the kidney required cortisol (Levi et al., 1973). Thus it is reasonable to postulate that the low plasma corticosterone levels reduce the inhibitory effect of NE on the action of ADH, which, in turn, decreases urine volume and conserves water.

Contrary to the finding of Vanjonack et al. (1975), plasma levels of 17-hydroxycorticosteroids were higher in the desert rodent *Jaculus jaculus* than in the semidesert rodent *Gerbillus gerbillus* (Haggag and El-Husseini, 1974). This discrepancy deserves some consideration. First, the method used by Haggag and El-Husseini to determine plasma glucocorticoids has several limitations, as discussed by Noujaim and Jeffery (1970). In my opinion, *J. jaculus* and *G. gerbillus* probably do not secrete 17-hydroxycorticosteroids (cortisol or hydrocortisone), but may secrete primarily corticosterone as most rodents do. The primary glucocorticoid secreted by the two Egyptian rodents should be determined. Second, the postulate that high levels of 17-hydroxycorticosteroids are necessary to maintain the rise in body temperature which results in storage of heat in the body rather than the use of water for its dissipation (Haggag and El-Husseini, 1974) is unacceptable. Storage of heat, i.e., hyperthermia, to conserve water has some advantage for a large animal, but its advantage is doubtful for a small animal. For a rodent weighing 100 g, the storage of an insignificant amount of heat would result in an untolerable rise in body temperature. To explain an increase in the secretory function of the adrenal cortex of desert rodents in terms of heat storage and, in turn, water conservation is an oversimplification. Third, Haggag and El-Husseini (1974) have suggested that an increase in plasma 17-hydroxycorticosteroids protects the animal indirectly from water loss. This hypothesis is unacceptable in light of evidence presented earlier on the interrelationship of glucocorticoids, NE, and ADH. It is difficult at this time to offer a reasonable explanation for the high plasma levels of glucocorticoids seen in *J. jaculus* and *G. gerbillus* reported by Haggag and El-Husseini (1974) and in *A. leucurus* reported by Vanjonack et al. (1975).

CONCLUDING REMARKS

The role of the endocrine system in adaptations of desert mammals is becoming better understood. Although methodology used in the study of secretory activity of endocrine glands has advanced greatly in the last decade, insufficient effort has been exerted to elucidate the endocrine adjustments that enable mammals to survive in the desert. Despite the inadequacies of available evidence, there seem to be significant differences associated with ecologic distribution in the secretory activity of the thyroid and adrenal cortex glands of various rodents. These hormonal differences appear to coincide well with the different MR acquired by the various species. Comprehensive reviews on MR of desert rodents have been published recently (Bradley and Yousef, 1972; Hart, 1971). The evidence is fairly convincing that the low resting MR of desert rodents is of adaptive significance for survival. Available data on thyroid and adrenal cortical secretory activities seem to offer, at least in part, a reasonable mechanism to explain the low MR of desert rodents. Before one can draw a convincing picture of the regulatory role of the endocrine system in adaptation of desert animals, a great deal of research in this area is required. It is hoped that this presentation will stimulate desert physiologists to explore the integrative role of the pituitary, thyroid, parathyroid, and adrenal glands in the mechanisms necessary for heat production, heat loss, and water and electrolyte balance in desert organisms.

LITERATURE CITED

Baird, D. T., R. Horton, C. Longcope, and J. F. Tait. 1969. Steroid dynamics under steady-state conditions. Rec. Progr. Hormone Res. 25:611.

Benedict, F. G. 1938. Vital Energetics. Carnegie Inst. Wash. Publ. 503.

Blahd, W. H. 1971. Radioisotope hormone assay methods. In Nuclear medicine, 2nd ed., p. 620. McGraw-Hill, New York.

Bradley, W. G., and M. K. Yousef. 1972. Small mammals in the desert. In M. K. Yousef, S. M. Horvath, and R. W. Bullard (eds.), Physiological adaptations: desert and mountain, p. 127. Academic Press, New York.

Brody, S. 1945. Bioenergetics and growth. Reinhold, New York.

———, and R. C. Procter. 1932. Growth and development with special reference to domestic animals. Further investigations of surface area in energy metabolism. Missouri Agr. Exp. Sta. Res. Bull. 116.

Brown, G. W. (ed.). 1968. Desert biology, Vol. 1. Academic Press, New York.

Burkle, J. S., and R. Lund. 1963. Measurements of thyroid function with various radioisotopes of iodine. In F. W. Sunderman and F. W. Sunderman, Jr. (eds.), Evaluation of thyroid and parathyroid functions. J. B. Lippincott Co., Philadelphia.

Cassuto, Y., and Y. Amit. 1968. Thyroxine and norepinephrine effects on the metabolic rates of heat acclimated hamsters. Endocr. 82:17.

Chaffee, R. R. J., and J. C. Roberts. 1971. Temperature acclimation in birds and mammals. Ann. Rev. Physiol. 33:155.

Chattoraj, S. C. 1970. Endocrinology. In N. W. Tietz (ed.), Fundamentals of clinical chemistry, p. 474. W. B. Saunders Co., Philadelphia.

Collins, K. J., and J. S. Weiner. 1968. Endocrinological aspects of exposure to high environmental temperatures. Physiol. Rev. 48:785.

Dill, D. B., E. F. Adolph, and D. G. Wilber (eds.). 1964. Adaptation to environment. Handbook of Physiology, Sec. 4, American Physiological Society, Washington, D.C.

Ekins, R. P., E. S. Williams, and S. Ellis. 1969. The sensitive and precise measurement of serum thyroxine by saturation analysis (Competitive protein binding assay). Clin. Biochem. 2:253.

Eleftheriou, B. E. 1964. Bound and free corticosteroid in the plasma of two subspecies of deer mice after exposure to a low ambient temperature. J. Endocr. 31:75.

Evans, E. S., A. N. Contopoulos, and M. E. Simpson. 1957. Hormonal factors influencing calorigenesis. Endocr. 63:836.

Folk, G. E., Jr. 1974. Textbook of environmental physiology, 2nd ed. Lea and Febiger, Philadelphia.

Fregley, M. J., K. M. Cook, and A. B. Otis. 1963. Effect of hypothyroidism on tolerance of rats to heat. Amer. J. Physiol. 204:1039.

Gross, H. A., H. J. Puder, K. S. Brown, and M. B. Lipsett. 1972. A radioimmunoassay for plasma corticosterone. Steroids 20:681.

Guilleman, R., G. W. Clayton, H. S. Lipscomb, and J. D. Smith. 1959. Fluorometric measurement of rat plasma and adrenal corticosterone concentration. J. Lab. Clin. Med. 53:830.

Haggag, G., and M. El-Husseini. 1974. The adrenal cortex and water conservation in desert rodents. Comp. Biochem. Physiol. 47:351.

Hart, J. W. 1971. Rodents. In G. C. Whittow (ed.), Comparative physiology of thermoregulation, Vol. 2. Academic Press, New York.

Hayward, J. S. 1965. Metabolic rate and its temperature—adaptive significance in six geographic races of *Peromyscus*. Can. J. Zool. 43:132.

Heroux, O., and R. Brauer. 1965. Critical studies on determination of thyroid secretion rate in cold-adapted animals. J. Appl. Physiol. 20:597.

Hoff, C. C., and M. L. Riedesel (eds.). 1969. Physiological systems in semi-arid environments. University of New Mexico Press, Albuquerque, N.M.

Hoffman, R. A. 1964. Terrestrial animals in cold: hibernators. In D. B. Dill, E. F. Adolph, and C. G. Wilber (eds.), Adaptation to the environment, Handbook of Physiology, Sec. 4, p. 379. American Physiological Society, Washington, D.C.

Hudson, J. W. 1964. Temperature regulation in the round-tailed ground squirrel, *Citellus tereticaudus*. Ann. Acad. Sci. Fennicae, AIV, 71:16.

————, and L. C. Wang. 1969. Thyroid function in desert ground squirrels. In C. C. Hoff and M. L. Riedesel (eds.), Physiological systems in semi-arid environments. University of New Mexico Press, Albuquerque, N.M.

Huibreytse, W. H., R. Gunville, and F. Ungar. 1971. Secretion of corticosterone *in vitro* by normothermic and hibernating ground squirrels. Comp. Biochem. Physiol. 38A:763.

Johnson, H. D. 1972. Environmental temperature effects and hormonal control on heat production of cattle. In Isotope studies on the physiology of domestic animals. International Atomic Energy Agency, Vienna.

————. 1974. Tracer studies in environmental adaptation. In Tracer techniques in tropical animal production. International Atomic Energy Agency, Vienna.

Kase, N. Advances in steroid assays. 1970. In S. H. Sturgis and M. L. Taymor (eds.), Progress in gynecology, Vol. V, p. 119. Grune & Stratton.

Kleiber, M. 1932. Body size and metabolism. Hilgardia 6:315.

————. 1961. The fire of life. John Wiley & Sons, New York.

Lemaine, R., O. Olsen, and C. Benceny. 1965. Le role des hormones cortico-surrenal dans l'adaptation a la chaleur. Ach. Sci. Physiol. 19:141.

Levi, J., S. G. Massry, and C. R. Kleeman. 1973. The requirement of cortisol for the inhibitory effect of norepinephrine on the antidiuretic action of vaso-pressin. Proc. Soc. Exp. Biol. Med. 142:687.

Luick, J. R. 1970. Use of radioactive isotopes in veterinary clinical biochemistry. In J. J. Kaneko (ed.), Clinical biochemistry of domestic animals, 2nd ed., Vol. 2, p. 271. Academic Press, New York.

Maloiy, G. M. O. (ed.). 1972. Comparative physiology of desert animals. Symposia of the Zoological Society of London, No. 31. Academic Press, New York.

McNab, B. K. 1968. The influence of fat deposits on the basal rate of metabolism in desert homoiotherms. Comp. Biochem. Physiol. 26:337.

Midgley, A. R., Jr., G. D. Niswender, V. L. Gay, and L. E. Reichert, Jr. 1971. Use of antibodies for characterization of gonadotropins and steroids. Rec. Progr. Hormone Res. 27:235.

Moran, J. J. 1952. Factors affecting the determination of protein-bound iodine in serum. Anal. Chem. 24:378.

Murphy, B. E. P., C. J. Pattee, and A. Gold. 1966. Clinical evaluation of a new method for the determination of serum thyroxine. J. Clin. Endocr. 26:247.

Noujaim, A. A., and D. A. Jeffery. 1970. Analysis of corticosteroids in biological samples. I. Spectrophotometric, fluorometric and chromatographic methods. Can J. Pharmaceutical Sci. 5:26.

Odell, W. D., and W. H. Daughaday (eds.). 1972. Principles of competitive protein-binding assays. J. B. Lippincott Co., Philadelphia.

Popovic, V. Endocrines in hibernation. 1960. In C. P. Lyman and A. R. Dawe (eds.), Mammalian hibernation, p. 105. Mus. Comp. Zool. Bull. 124.

Powsner, E. R., and D. E. Raeside. 1971. Competitive binding, dilution derivative and other *in vitro* radioassays. In Diagnostic nuclear medicine, p. 558. Grune & Stratton, New York.

Robertshaw, D. (ed.). 1974. Environmental physiology. MTP International Review of Science, Physiology Series One, Vol. 7. University Park Press, Baltimore.

Rubin, M. 1970. Fluorometry and Phosphorimetry in clinical chemistry. Adv. Clin. Chem. 13:161.

Schmidt-Nielsen, K. 1964. Desert animals: Physiological problems of heat and water. Oxford University Press, New York.

Scholander, P. F., R. Hock, V. Walters, and L. Irving. 1950. Adaptation to cold in arctic and tropical mammals and birds in relation to body temperature, insulation and basal metabolic rate. Biol. Bull. 99:259.

Scott, I. M., M. K. Yousef, and W. G. Bradley. 1972. Body fat content and metabolic rate of rodents: desert and mountain. Proc. Soc. Exp. Biol. Med. 141:818.

Silber, R. H. 1966. Fluorometric analysis of corticoids, Vol. 14. In D. Glick (ed.), Methods in biochemical analysis, p. 63. Wiley-Interscience, New York.

———, R. D. Busch, and R. Oslapas. 1958. Practical procedure for estimation of corticosterone or hydrocortisone. Clin. Chem. 4:278.

Sparagana, M., G. Phillips, and L. Kucera. 1969. Serum thyroxine by competitive protein binding analysis: clinical, statistical and comparative evaluation. J. Clin. Endocr. 29:191.

Tashima, L. S. 1965. The effects of cold exposure and hibernation on thyroidal activity of mesocricetus auratus. Gen. Comp. Endocrin. 5:267.

Turner, C. W. 1969. Method of estimating thyroid hormone secretion rate of rats and factors affecting it. Missouri Agr. Exp. Sta. Res. Bull. 969.

Vanjonack, W. J., I. M. Scott, M. K. Yousef, and H. D. Johnson. 1975. Corticosterone plasma levels in desert rodents. Comp. Biochem. Physiol. (in press).

Whittow, G. C. 1971. Ungulates. In G. C. Whittow (ed.), Comparative physiology of thermoregulation, Vol. 2. Academic Press, New York.

Wilson, W. O., H. D. Johnson, and N. Pace. 1971. Physiologic functions and measurement techniques. In A guide to environmental research on animals, p. 93. National Academy of Sciences, Washington, D.C.

Yousef, M. K., R. D. Cameron, and J. R. Luick. 1971. Hydrocortisone secretion rate in reindeer, *Rangifer tarandus:* Effects of season. Comp. Biochem. Physiol. 40:495.

———, and D. B. Dill. 1971a. Responses of Merriam's kangaroo rats to heat. Physiol. Zool. 44:33.

———, and D. B. Dill. 1971b. Daily cycles of hibernation in the kangaroo rats, *Dipodomys merriami.* Cryobiology 3:122.

———, S. M. Horvath, and R. W. Bullard (eds.). 1972. Physiological adaptations: desert and mountain. Academic Press, New York.

———, and H. D. Johnson. 1966. Calorigenesis of dairy cattle as influenced by thyroxine and environmental temperature. J. Animal Sci. 25:150.

———, and H. D. Johnson. 1967a. Calorigenesis of cattle as influenced by hydrocortisone and environmental temperature. J. Animal Sci. 26:1087.

———, and H. D. Johnson. 1967b. Time course of oxygen consumption in rats during sudden exposure to high environmental temperature. Life Sci. 6:1121.

———, and H. D. Johnson. 1967c. A rapid method for estimation of thyroxine secretion rate of cattle. J. Animal Sci. 26:1108.

———, and H. D. Johnson. 1968. Effects of heat and feed restriction during growth on thyroxine secretion rate of male rats. Endocr. 82:353.

———, and H. D. Johnson. 1972. Thyroid function and metabolic rate of desert rodents. In S. W. Tromp and J. J. Bouma (eds.), Biometeorology, Vol. 5(1), p. 134. Swets Zeitlinger, N. V., Amsterdam.

———, H. D. Johnson, W. G. Bradley, and S. M. Sief. 1974. Tritiated water turnover rate in rodents: desert and mountain. Physiol. Zool. 47:153.

———, and J. R. Luick. 1971. Estimation of thyroxine secretion rate in reindeer, *Rangifer tarandus:* effects of sex, age and season. Comp. Biochem. Physiol. 40:789.

———, D. Robertson, and H. D. Johnson. 1967. Effect of hibernation on oxygen consumption and thyroidal I[131] release rate of *Mesocricetus auratus.* Life Sci. 6:1185.

———, Y. Takahashi, W. D. Robertson, L. J. Machlin, and H. D. Johnson. 1969. Estimation of growth hormone secretion rate in cattle. J. Animal Sci. 29:341.

ADAPTATIONAL BIOLOGY OF DESERT TEMPORARY-POND INHABITANTS

Denton Belk and Gerald A. Cole

Department of Zoology, Arizona State University, Tempe, Arizona

Abstract

Animals inhabiting ephemeral waters, regardless of regional distribution, are adapted for living in a chemically and physically unstable medium, making a safe transition between dry and wet conditions, and surviving dry phases. Desert species are unique only in their capacity to adjust to the driest, hottest, and most rapidly changing conditions, and the irregularity of precipitation adequate to fill ponds.

Crustacean inhabitants survive dry phases in various resting stages, most commonly as desiccated and probably ametabolic embryos. At least one fairy shrimp, *Streptocephalus seali,* is apparently unable to colonize desert ponds because its embryonic tissues are intolerant to prolonged severe drying. Some insect larvae, such as *Polypedilum vanderplanki,* also await the return of water in a dehydrated, ametabolic state. Other insects mature rapidly and move from the drying pond to more permanent aquatic situations.

Reactivation of dormant resting stages is typically keyed to a combination of factors that indicate reestablishment of an aquatic phase under conditions favorable to growth and reproduction. Temperature is fundamental in all cases. Other cues include changes in salinity, oxygen concentration, light intensity, and photoperiod. The effect of these stimuli are influenced by past history of the dormant stage.

Desert and nondesert crustaceans acclimate rapidly to gross physical and chemical instabilities associated with temporary ponds. Several species tolerate decreased oxygen concentrations reaching less than 1 ppm. They also employ behavioral adaptations, actively moving into pond strata where thermal and oxygen conditions are most favorable.

INTRODUCTION

Inhabitants of temporary aquatic environments must possess adaptations for living in a chemically and physically unstable medium, making a safe transition between wet and dry conditions, and surviving during dry phases. The overall picture is complex, but *synchrony* is the key word; life cycles of the organisms must mesh with changing conditions of their habitats.

The typical temporary pond is a shallow, closed basin. Such habitats occur in both desert regions and in climates where the mean annual precipitation exceeds the yearly evaporation rate. Regardless of their location, ephemeral waters may be grouped into two main categories according to Decksbach (1929; translated by Hartland-Rowe, 1972): (1) seasonally astatic waters that dry up annually, and (2) perennially astatic waters that show striking annual fluctuations in level, but do not disappear every year.

Some seasonally astatic ponds have a nearly regular regime, with the aqueous phase occurring at approximately the same time each year. These are typified by vernal pools in northern or high-altitude regions where melting snow is a regular event and serves as an unfailing source of water. One type of vernal pond is essentially permanent, but during winter months it is frozen throughout, the ice producing a "physiologic" dry phase. Many shallow arctic and alpine lakes are of this type and support a temporary-pond fauna during the summer.

A regular periodicity may also exist for transitory waters in warm, humid regions where flooding depends on predictable annual patterns of precipitation. The Louisiana ponds investigated for many years by Moore and his students are examples of this type (see Moore, 1970).

In the deserts of the world, seasonally astatic ponds are characterized by irregularity of appearance. For example, in the Sonoran Desert of Arizona, ponds usually fill with warm water during summer rains, dry, and refill with colder water from winter storms. There are years, however, when a given pond may fill during only one of these wet seasons. Depressions in a desert often will be dry throughout the year or for several years. In some instances, the basin may not fill for decades, as is the case with Lake Eyre in Australia (Bayly and Williams, 1973). Lack of periodicity is further evidenced in deserts by fillings and dryings within a single season.

Although reference will be made to temporary-pond animals from alpine, northern, and warm humid regions, this discussion is focused on

:he fauna of arid, hot deserts. Consideration of species from the nondesert areas is justifiable, since many adaptations to astatic ponds are
shared by members of the ephemeral water fauna regardless of regional
distribution. Major differences are the degree to which desert forms
withstand the driest, hottest, and most rapidly changing conditions, and
the irregularity of adequate precipitation to fill the ponds.

ADAPTATIONS FOR SURVIVAL DURING
THE DRY PHASE

Resting Stages

Clearly defined resting stages are common in the life cycles of organisms that inhabit unstable environments. These are periods of dormancy when metabolism is greatly reduced, and growth and reproduction cease. There is also an enhanced ability to survive temperature
extremes, anoxia, and desiccation. In most species only specific stages
in the life cycle are capable of undergoing dormancy and arrested
development.

The different types of dormancy have been classified in several ways;
the categories have been reviewed by Lees (1955) and Sussman and
Halvorson (1966). Some authors have stressed a scale of metabolic
activity ranging from 100 percent, as in active life, to 0 percent when
there is no perceptible metabolism, the latter state being aptly termed
cryptobiosis by Keilin (1959). Other classifications overlap and hinge on
whether dormancy occurs because of an innate property of the organism with little effect from environmental factors (*constitutive dormancy*), or because of the immediate effects of unfavorable physicochemical conditions of the environment (*exogenous dormancy*) (Sussman and Halvorson, 1966, Table 1.1).

Just as a continuous spectrum exists between vigorously metaboliz-
ing organisms and those in which no metabolism can be perceived, so
there are blurred distinctions between the categories of constitutive and
exogenous dormancy. Unfortunately, arbitrary definitions have removed some serviceable words from general usage. *Quiescence,* for
example, is an exogenous resting period brought on immediately by
unfavorable conditions. *Diapause* is an example of a constitutive arrest
of development initiated by internal factors before the environment has
become unfavorable. *Obligatory diapause* occurs every generation, and
seems to be spontaneous and indifferent to environmental cues. *Facultative diapause* does not occur every generation and appears to be
evoked by environmental factors that predict a change to unfavorable
conditions. Its similarity to quiescence is obvious, and the importance
of this sort of developmental arrest in synchronizing the life cycles to

habitat instabilities is evident. To emphasize further the overlap in definitions, obligatory diapause may not be entirely immune to environmental stimuli, but might be a response to such a broad spectrum of cues that it appears in every generation. This makes it difficult to sort out the responsible factor or factors of the milieu and lends strength to the concept of an inherent spontaneity initiating and terminating diapause.

From an ecological point of view, obligate diapause would be the most appropriate strategy in predictably cyclic environments, whereas quiescence and facultative diapause would be optimal in less certain situations.

States of apparent dormancy can include morphogenesis in addition to quiescence and diapause. Only by experimental study can one understand the situation for any given organism. For example, Broch (1965) found that seemingly dormant eggs of the fairy shrimp *Eubranchipus bundyi* developed from stage 1 to stage 3 embryos during the first 12 days following expulsion from the ovisac as long as they were in aerobic water around 20 to 30°C. Having reached stage 3, these embryos became quiescent unless moved to 7°C, at which temperature they continued development. He also found that the recently dropped eggs entered quiescence if their environment was cold (7°C) or anoxic.

According to Keilin (1959), the conditions that can bring about cryptobiosis, either singly or in combination, are loss of water, lowering of temperature, absence of oxygen, and high salt concentration. It seems to us that high temperature may need to be added to this list. The energy savings resulting from ametabolic existence would be highly adaptive in situations such as that found in *Eubranchipus bundyi,* where embryos become quiescent after reaching stage 3 when the temperature stays around 20°C. This possibility should be studied.

Survival as Eggs

A common stage in surviving the arid phase of a desert pond is the "egg" of the anostracans (fairy shrimps), conchostracans (clam shrimps), and notostracans (tadpole shrimps), collectively referred to as phyllopods. The egg is actually a shelled embryo, dormant during most of its existence.

Clegg (1967) showed that the encysted embryos of *Artemia salina* enter cryptobiosis when they desiccate as a normal and necessary part of their development. The embryonic tissues of two conchostracan species were observed to survive dehydration, shrinking to shiny, orange, cup-like masses occupying only a fraction of the space enclosed by the tertiary shells (Bishop, 1968; Belk, 1970). Cryptobiosis undoubtedly occurred in these young clam shrimps, too. It is highly probable that this ametabolic state is a common feature of phyllopod eggs, and functions to conserve the limited resources stored within these small structures that are typically subjected to long periods of dormancy.

The thick, spongy tertiary shell of phyllopod eggs is probably a protective structure. As indicated, it is not a watertight covering that keeps the contents from dehydrating. Indeed, such a situation would hardly be functional in light of the osmotic nature of the hatching mechanism in these animals (Clegg, 1964; Broch, 1965; Belk, 1972). Belk (1970) demonstrated that even though deshelling embryos of *Eulimnadia antlei* (Conchostraca) did not harm them, they were unable to withstand shaking in sand or exposure to strong sunlight in the deshelled condition. Broch (1965) observed fungus frequently killed embryos of *Eubranchipus bundyi* (Anostraca) after he removed their shells.

In view of the protective nature of the tertiary shell, it is interesting that not all clam shrimps from desert regions have such structures. *Leptestheria compleximanus,* which inhabits ephemeral waters in the Sonoran Desert of Arizona, is an example. Nothing is known about how such eggs withstand harsh desert conditions or how they compare in durability to shelled forms produced by the more common Sonoran Desert conchostracan, *Eulimnadia texana.*

In another group of crustaceans, the Cladocera, dormant resting eggs are enclosed in a specially developed part of the carapace known as an ephippium. Like the phyllopod tertiary shell, the ephippium is not watertight; thus, ability to survive desiccation, a characteristic of some species, must be attributed to the embryonic tissue. The role of the ephippial case seems to be protective like that of the phyllopod tertiary covering. Shan (1969) suggested that the case shelters the embryo from invertebrate predators and harmful effects of light.

The two unusual genera of chydorid cladocerans, *Eurycercus* and *Saycia,* were recently reviewed by Frey (1971). He pointed out that *Saycia,* known only from Australia and New Zealand, is restricted to temporary waters, and it has a better developed ephippium than *Eurycercus.* The function of this egg case in the adaptational biology of desert ephemeral water cladocerans deserves study.

A few species of calanoid copepods are adapted for occupancy of temporary ponds. In well-watered regions certain species are found only in temporary sites; in the Arizona desert several species occur in both temporary and permanent habitats (Cole, 1966). This is true also for *Boeckella triarticulata,* the commonest Australian species of the genus (Bayly, 1964; Bayly and Williams, 1973). In the diaptomids at least, males are common in all instances, although sex ratios vary. Apparently, all eggs are fertilized, but there are two kinds. One type, the subitaneous egg, is thin shelled and usually hatches immediately; the second type is a resting egg that drops from the female and releases the nauplius after a prolonged dormancy, including diapause. The two types can be differentiated by the color and nature of the thicker shell in the latter. *Diaptomus stagnalis* (Brewer, 1964; Sawchyn and Hammer, 1968), *D. kiseri, D. arcticus,* and *D. sanguineus* (Sawchyn and Hammer, 1968) produce only the resting egg. *Diaptomus lilljeborgi* of North Africa (Gauthier, 1928) and *D. clavipes* of the American Southwest produce both types (Gehrs and Martin, 1974). The last two species occur in both temporary and permanent habitats. *Diaptomus siciloides,* common in permanent lakes of North America, produces both subitaneous and overwintering eggs that survive in the sediments underlying the water (Comita, 1972). These delayed-hatching eggs must be able to tolerate some desiccation as this species is fairly common in Arizona temporary ponds (Cole, 1966).

In certain instances subitaneous eggs are said to be capable of resting. Gurney (1931) wrote of a calanoid population, derived from what appeared to be subitaneous eggs, that hatched when a temporary pond refilled. O'Brien et al. (1973) described a population of *Diaptomus leptopus* that produced only subitaneous eggs in a temporary pond. Fairbridge (1945) reported the same for *Boeckella opaqua* in ephemeral, rain-filled depressions in granite rocks of Western Australia.

It seems likely that the thick shell of the calanoid copepod egg functions much like the protective coverings discussed for phyllopods and cladocerans. The occurrence of both thick-shelled and subitaneous resting eggs in temporary-pond forms recalls the situation mentioned earlier in the Sonoran Desert for the conchostracans *Leptestheria compleximanus* and *Eulimnadia texana.* Both observations suggest that shells play only a secondary role in the adaptation of crustacean eggs to desiccating conditions.

Survival as Immature or Adult Stages

Among at least 20 species of cyclopoid copepods, an arrested development that seems to fit the definition of diapause occurs in immature copepodid stages (Elgmork, 1967; Champeau, 1966). The most remarkable of these was reported by Rzóska (1961). In depressions that hold temporary rainpools near Khartoum, *Metacyclops minutus* survives the hot Sudanese sun as resting copepodids for 9 months before emerging. Although *Acanthocyclops vernalis* and *Eucyclops agilis* occur in some Arizona transitory ponds (Cole, 1968) and cyclopoids occur in Australian rainpools (Bayly and Williams, 1973), the mechanisms for surviving the dry phase are not known.

The first example of a cyclopoid that encysts and diapauses, *Cyclops bicuspidatus thomasi,* was discovered by Birge and Juday (1908). It is a common planktonic species in North America, where it often diapauses in the fourth copepodid stage (Cole, 1953). The resting copepodids are incapable of withstanding desiccation; the cyst is an incomplete gelatinous envelope. O'Brien et al. (1973) reported it from a temporary pond in Ontario, a finding which suggests that it might survive by diapausing in moist soils.

Some harpacticoid copepods encyst as morphologic adults (Deevey, 1941; Fryer and Smyly, 1954) in permanent waters. A few species were reported from temporary waters in Africa (Gauthier, 1928), and survival at the adult stage was implied. Champeau (1967) has supplied further details about the problem. *Cletocamptus retrogressus* persists for 6 months in the sediments of temporary saline ponds in Europe as adult females. The immature copepodid stages and males are capable of resting for only short periods, and so the adult females with spermatozoa in their seminal receptacles are the aestival quiescent stage. When revived at the pond's refilling, they produce several successive egg sacs in which the eggs develop normally to start the new generation.

Dussart (1967) hinted at an exception to the idea that temporary pond diaptomids can survive only in the egg stage. He writes (p. 56) that some preliminary observations on temporary waters in the south of France seem to show that *Diaptomus cyaneus* is capable of surviving the dry period of its habitat in the fourth copepodid stage. If true, this is a cyclopoid feature. The species is not found in the most arid regions. Gauthier (1928) showed that it occurs in the temporary ponds of what he designated the Algerian pluvial zone, where the annual precipitation is about 50 cm; but it is replaced by *D. chevreuxi* in the *zone steppique,* where the precipitation is only 30 cm/year.

Edward (1964, 1968) reported that two chironomid larvae (*Allotrissocladius amphibius* and *Paraborniella tonnoiri*) form capsules within which they survive during the dry phase of certain Australian rockpools. These desiccated basins can heat to from 56 to 60°C for up to 2 hours per day. The larvae remain hydrated during this dormant period although they may become quite flaccid, indicating considerable loss of body water. Some individuals of *A. amphibius* were observed to survive apparent total dehydration, a situation suggesting that the adaptational biology of this species may be comparable to the cryptobiotic abilities of *Polypedilum vanderplanki.* In this connection, it should be noted that Hinton (1960) considers the capacity to tolerate a high degree of dehydration reasonably certain evidence of cryptobiosis.

The larvae of a dipteran assignable to the genus *Dasyhelea* in the family Ceratopogonidae (= Heleidae) also survive desiccation. Tribbey (1965) found these living in rock depression ponds atop Enchanted Rock in Texas.

How Dry?

Aridity of the dry phase varies. Moore (1967) found the soil moisture of temporary ponds in Louisiana ranged from 60 percent of the dry weight to a low of 9.5 percent. The sunbaked soils of dry pools in desert regions must have far lower water content for longer periods.

Two studies suggest that degree of desiccation may limit the survival of phyllopod embryos and thus exclude all but the most tolerant species from deserts. First, working with the tadpole shrimp, *Triops granarius,* from the hot desert near Khartoum in Sudan, Carlisle (1968) found death occurred only under temperature and pressure conditions that boiled off the last small amount of water retained within the dry eggs. At local atmospheric pressure, resulting in a 100°C boiling point for water, experimental eggs hatched after 16 hours in an oven at 98°C, but failed to eclose after only 15 minutes at 102°C. In experimental containers where pressure was increased to raise the boiling point to 105°C, eggs survived 16 hours at 103°C. However, eggs failed to hatch after only 30 minutes at 75°C when reduced pressures lowered the boiling point to 70°C. In a second study, Moore (1967), working with the fairy shrimp *Streptocephalus seali* from Louisiana ponds, found that storing eggs for 60 to 90 days at 0 percent relative humidity significantly decreased survival. Only 5 of 400 hatched compared with 265 of 400 kept at 100 percent relative humidity.

Streptocephalus seali occurs on the high plateaus in Arizona, though it is absent from dryer desert sections of the state (Belk, 1974). The long periods of extreme aridity in the latter regions probably operate as the major limiting factor.

Roy (cited in Hutchinson, 1967) discussed experiments on the cyclopoid copepod *Cyclops furcifer.* He found that both eggs and last copepodites or adults could successfully undergo desiccation only when dried in a protective layer of mud. It may be that the dirt around them held a critical level of moisture, thus preventing total dehydration and death.

The question of how dry conditions must be in a given pond basin to affect the survival and thus occurrence of particular species is open to investigation. Information of this type is likely to clear up some puzzling patterns observed in the distribution of temporary pond organisms.

Escape to Refugia

In some ephemeral water species, the adults maintain themselves in permanent aquatic habitats, or elsewhere, and disperse rapidly to newly formed temporary habitats. Hynes (1955) reported an example of this for insects in East Africa. Six species of the Hemiptera survive in permanent pools until rains create temporary habitats, whereupon they fly up to 100 km across arid lands to colonize the new ponds, where breeding may occur. In Arizona, *Ranatra, Notonecta,* and some coleopterans find their way to temporary ponds to feed upon the crustaceans (Cole, 1968). Similarly, Sublette and Sublette (1967) reported predaceous insects migrating into Texas and New Mexico playas.

In many deserts, anurans such as spadefoot toads adjust their aquatic phase to the short duration of ephemeral ponds by undergoing comparatively rapid embryonic and larval development. Then as juveniles and adults they burrow to escape desiccation (Mayhew, 1968; Ruibal et al., 1969).

ADAPTATIONS TO THE DRY–WET INTERPHASE

Sorting out the environmental factors that stimulate emergence or hatching is often complicated. First, several factors frequently work together; second, the efficacy of a stimulus may depend on the past history of the dormant stage; third, the effectiveness of any environmental cue may vary among species and conspecific populations (i.e., physiological races).

As an example of the last point, Belk has tried hatching eggs of the fairy shrimp *Streptocephalus seali* from different localities. Eggs supplied by Walter G. Moore from Louisiana populations hatched readily. However, those collected from *S. seali* living on the Kaibab Plateau in Arizona repeatedly failed to eclose under identical laboratory conditions.

One can envision the effect of past history, the second complicating factor, by reference to Broch's (1965) data on embryonic development in *Eubranchipus bundyi.* If two eggs dropped by a female end up in different microhabitats toward the end of the warm season when the pond is drying, their subsequent development and time of hatching can be quite different. The embryo in a warm, oxygen-rich area will undergo morphogenesis to stage 3 and become quiescent. Given favorable conditions, this egg will give rise to a young anostracan the following spring. The second egg, on the other hand, will be kept dormant first by a lack of oxygen, next by desiccation, and finally by autumnal low temperatures, even if the egg is now in an oxygen-rich environment. This embryo will not be ready to emerge during the spring, when its sibling hatches. If, however, it lies in an oxygenated area of the basin during the second summer, development will proceed to a stage 3 embryo, and with appropriate conditions during the next fall and winter, this young fairy shrimp will be ready to eclose 1 year later than its sibling.

Eggs from a Sonoran Desert fairy shrimp, *Branchinecta lindahli,* hatch within 2 days at 15°C, but fail to eclose at 30°C (Belk, 1973, 1974). The high temperature simply blocks development; it does not harm the embryos. When eggs are incubated at 30°C for 12 hours and then moved directly to 15°C where they are kept for 4 days, fewer than 10 percent hatch. This implies that the embryos of this species remain dormant if conditions are not favorable to hatching during the first few hours following rehydration. Similar reactions may be widespread among desert species and probably function to limit eclosion to the early filling phase of the pond. Late-hatching phyllopods would face a shortened time in which to reach maturity, a factor that would act as a selective agent working against delayed emergence.

The major conditions that terminate dormancy are usually associated with sudden changes brought about by the pond's origin. Inundation itself is a momentous event, the addition of water being important to all species. In humid regions, resting stages may be continuously moist and, therefore, water's arrival is not as abrupt an environmental change as in dusty desert depressions.

For eggs that have lain dormant in aerobic surroundings, a sharp decline in oxygen tension soon after flooding may be the final factor that synchronizes hatching with pond formation. The sharp drop in oxygen seems to be a function of bacterial action and organic sediments in the new pond. It is the stimulus that causes hatching of *Diaptomus stagnalis* and *D. caducus* in the northern United States and Canada (Brewer, 1964). It effects the final emergence of the metanauplii of *Eubranchipus bundyi* in woodland pools of New York State (Broch, 1965), and it breaks diapause in the eggs of *Aëdes* mosquitoes (Horsfall, 1956) and in the fully developed embryos of some cyprinodont fishes found in temporary ponds (Peters, 1963). However, Brown and Carpelan (1971) questioned the importance of this factor in desert situations.

In desert pools lying above saline substrates, the sudden arrival of dilute water at the time of filling induces hatching. Brown and Carpelan (1971) showed that *Branchinecta mackini* eggs hatch at the initial flooding of Rabbit Dry Lake, but found that rapidly mounting salinity prohibits further eclosion. The environmental shock caused by sudden dilution from subsequent rains or melting snow causes more eggs to hatch. These later emergences are, of course, neatly synchronized with an event that portends prolongation of the playa's life.

Brown and Carpelan also suggested that salinity changes are important in waters of arid lands, but are of little consequence in humid areas where transitory waters are dilute. This conclusion may have been reached because many authors have not mentioned salinity in their papers " . . . justifiably considering it to be a factor of no concern" (Brown and Carpelan, 1971, p. 54).

Osmotic hatching mechanisms have been reported for a great many freshwater invertebrates (Davis, 1967), including phyllopods from dilute temporary ponds (Mattox, 1950; Broch, 1965; Belk, 1972) and from hypersaline waters (Clegg, 1964). Typically, the procedure for hatching freshwater phyllopod eggs is to place them in deionized or distilled water (Moore, 1967; Belk, 1972). The eggs may have been previously dried or, more significantly, they may have been stored for some time in habitat water before being placed in the distilled water. The arrival of dilute water probably represents an important salinity change in all temporary ponds, although it is less pronounced in humid than in arid regions, where substrates may be high in salts. Despite this qualification, the initial filling of all temporary ponds probably produces an osmotic shock that interacts with other environmental cues to trigger final hatching. In discussing eclosion in phyllopods, Hutchinson (1967, p. 562) went so far as to write, "It is quite likely that the fundamental change needed to produce hatching is always a lowering of osmotic pressure at the egg surface."

Several copepods from astatic saline ponds in Europe were studied by Champeau (1966), who reported that quiescence and reactivation can be provoked experimentally by varying chlorinity, but only for the proper stages—copepodids IV and V for cyclopoids, and adult females for *Cletocamptus retrogessus,* a harpacticoid. Later, Champeau (1967) showed that temperature, chlorinity, and desiccation interact to induce both inactivation and breaking of diapause in *Cletocamptus.* The rising temperature and concentration of solutes in evaporating waters lead females to burrow into the sediments rather than to follow the retreating shoreline.

A few studies have examined the role of light in initiating or terminating resting stages. Without convincing supporting data, Bishop (1967a) suggested that a photoperiod of 9–12 hours of light per day at the start of the Southern Hemisphere winter induces diapause in Australian conchostracans. By contrast, Horne (1971) found that three phyllopods from the American Southwest do not respond to day length.

Absence of light apparently inhibits hatching of some species. Pancella and Stross (1963) observed this in a cladoceran, and Bishop (1967b) and Belk (1972) recorded such an inhibiting effect upon two conchostracans. The last two authors suggested that darkness could be an important factor in keeping buried eggs from eclosing. This would prevent embryos covered by sediment from emerging and, therefore, the entire clam shrimp population would not hatch during a single filling of a pond.

Temperature exerts a controlling influence that is fundamental. If temperature conditions are favorable, the arrival of water in a new pond interacts with other factors to stimulate eclosion or to activate other resting stages. If not, hatching and activation will not occur.

Seasonal and geographic occurrence of active phyllopod populations is strongly influenced by temperature (Prophet, 1963; Horne, 1967; Moore, 1967; Belk and Belk, in press). Horne (1971) observed that three phyllopods, typically restricted to the warmer seasons in central Texas, hatched during a period of unusually high temperatures in January. Belk (1974) presented data on the hatching response in relation to a range of thermal conditions for several fairy shrimps. At constant temperatures, eggs of *Branchinecta lindahli*, a winter species in the Arizona Sonoran Desert, eclosed optimally at 5 to 20°C with a significant drop to around 10 percent at 25°C; no larvae emerged at 30°C. A common summer species in many of the same ponds, *Streptocephalus mackini*, eclosed optimally at around 25 to 30°C with no hatching at 10°C, and with a few eggs hatching at 42°C.

As discussed earlier, temperature controls periods of embryogenesis and dormancy in *Eubranchipus bundyi* (Broch, 1965). Brewer (1964) reported a similar situation for a calanoid copepod from a northern temporary pond.

The environmental cues used to stimulate emergence are typically those which best predict chances for favorable growth and reproduction in the pond concerned. Despite this, larvae are frequently "fooled" into emerging when insufficient water has entered the basin to allow them to reach maturity (Moore, 1955a). Phyllopods probably adapt to this by producing large numbers of eggs, which they scatter into different microhabitats of the pool where all of them will not experience favorable conditions for eclosion at any single flooding. Fairy shrimps from less predictable rain-filled ponds were shown by Belk (1974) to produce larger clutches than species from snow-melt pools that typically receive adequate water for their phyllopod populations to complete life cycles. Similarly, there are reports of more eggs per clutch in temporary ponds, based on comparisons made between conspecific calanoid copepods from permanent and impermanent waters (Røen, 1955; Cole, 1966). An adjustment to density-independent mortality seems to be the rule in temporary pond crustaceans; they appear to be *r*-selected, as recently elaborated by Pianka (1972) and Gadgil and Solbrig (1972).

ADAPTATIONS FOR SURVIVAL DURING THE WET PHASE

Desert aquatic habitats are characterized by diversity and markedly influenced by climatic fluctuations (Cole, 1968). This is especially true of ephemeral ponds.

Occupants of temporary ponds are often subjected to gross physicochemical instabilities. For example, Brown and Carpelan (1971) observed a salinity increase due to evaporation of 1213 ppm in a single day at Rabbit Dry Lake, California. Temperature has been observed to increase by 5°C in 1 hour (Hillyard and Vinegar, 1972) and range as much as 18°C during 1 day (Hartland-Rowe, 1972). These values are impressive when compared with the situation in permanent lakes.

Coping with and Using Thermal Conditions

The desert fairy shrimp *Streptocephalus mackini* can adjust rapidly to changing thermal conditions. When transferred directly from 11.5 to 25°C, a group of experimental *S. mackini* accomplished 48 percent of their acclimation to the latter temperature within 30 minutes, as judged by LD_{50} studies (Belk, 1974). Grainger (1958) observed that *Artemia salina* had similar abilities.

Other experiments have shown that individuals of several fairy shrimp species can recover from as much as 1 hour exposure to near-lethal high temperatures after being returned to more normal thermal conditions (Moore, 1955b; Belk, 1974). However, not all these species are desert forms. One, *Branchinecta paludosa,* is a common arctic-alpine fairy shrimp. Thus, ability to survive brief exposure to high temperature should be considered an adjustment to life in temporary ponds in general and not exclusively to desert waters. In this connection, it is worth noting that White (cited in Hartland-Rowe, 1972) found that daily temperature maxima lasted no more than 1 hour in a turbid pool in western Canada. The highest temperature he observed was 32°C.

A variety of thermal habitats are frequently available within temporary ponds. In two shallow, turbid puddles, temperature differences between surface and bottom waters ranged from 9 to 16°C (Eriksen, 1966). Cloudsley-Thompson (1966) demonstrated that two desert phyllopods can choose the most favorable of available temperature conditions.

Fast maturation is extremely important in short-lived desert ponds. Hillyard and Vinegar (1972) suggested that larval *Triops longicaudatus* (notostracan) and *Thamnocephalus platyurus* (anostracan) take advantage of periods of high temperature by greatly increasing their respiration and presumably their growth rate. These two workers measured changes in oxygen consumption between 26 and 30°C and found Q_{10} values of 4.12 for *Triops* and 11.81 for *Thamnocephalus.*

Coping with Low Oxygen

Temporary ponds in both desert and nondesert areas frequently have marked diel cycles in oxygen concentration as a result of the interaction of temperature and community metabolism (Hartland-Rowe, 1972). At times, dissolved oxygen may fall to extremely low levels. Studying two ephemeral ponds in central Texas, Horne (1971) observed on a hot day that oxygen dropped at night and remained below 1 ppm for as long as 8 hours. Moore and Burn (1968) recorded no subsurface oxygen for an 8-day period in a Louisiana forest pond.

Phyllopods cope with these stressful conditions partly through physiological tolerance. For example, Horne (1971) tested *Caenestheriella setosa* (conchostracan), *Triops longicaudatus,* and two anostracans (*Streptocephalus texanus* and *Branchinecta packardi*), and found that 50 percent survived 2 hours in 20 to 30°C water at 0.1, 0.55 to 0.65, and 0.75 to 0.85 ppm of oxygen, respectively. Moore and Burn (1968) likewise observed 2-hour LD_{50} values below 1 ppm of oxygen for *Eulimnadia inflecta* (conchostracan) and *Streptocephalus seali.*

Phyllopods also employ behavioral mechanisms to deal with the stress of oxygen depletion. Horne (1971) noted that the low 2-hour LD_{50} for *S. texanus* was reduced from 0.75 to 0.55 ppm of oxygen when the experimental animals could swim near an air–water interface. During the unusual 8 days of anoxic conditions in the subsurface waters of a Louisiana ephemeral pond, Moore and Burn (1968) found that *S. seali* survived by staying in the upper 2 cm where a small amount of oxygen was available. The heavy-bodied, weakly swimming conchostracan (*E. inflecta*) was unable to utilize the surface stratum and no individuals of this species survived.

Streptocephalus seali in Louisiana forest ponds is not a desert species. Thus, tolerance of low dissolved oxygen and swimming at the surface are not specific adaptations to desert ponds, but to the special oxygen relations of warm temporary ponds in both humid and arid regions.

In some species blood pigments play a part in coping with decreased oxygen supplies. Horne and Beyenbach (1971) found that the oxygen affinity of hemoglobin in *T. longicaudatus* has a relatively low sensitivity to increasing temperature. They theorized that this enhanced hemoglobin's role in assisting oxygen delivery to the tissues during periods of low dissoved oxygen resulting from high pond temperatures.

Coping with Salinity

Horne (1967) demonstrated that several essentially freshwater phyllopods can be acclimated slowly to approximately isosmotic NaCl solutions. Bayly (1972) pointed out that the upper salinity tolerance for many freshwater organisms occurs either somewhat before or immediately after their body fluids become isosmotic with the medium, and that experiments have disclosed the upper limit of survival is positively correlated with the initial concentration of the body fluid. Bayly offered two explanations for such an isosmotic limit: inability of the animal to obtain sufficient water from an isosmotic medium for excretory purposes, and the irreversible adaptation of the tissues of these animals to a dilute body fluid.

A common notostracan in ephemeral desert ponds in North America is *T. longicaudatus.* It was one of the species that Horne (1967) acclimated to approximately isosmotic conditions. In later studies, Horne (1968) demonstrated that this tadpole shrimp can withstand instantaneous exposure to salinities from 1 to 87 mmol/liter (156 to 13,112 ppm), which indicates good adaptation to fluctuating salinities. In terms of its ionic regulatory abilities, however, he found it is unable to control magnesium adequately when environmental concentrations reach 14 mmol/liter, which Horne suggested may explain the absence of this species from some otherwise suitable ponds.

Some animals are able to behave as osmotic conformers. Others can regulate hypoosmotically. These contrasting physiological abilities may have ecologic effects, resulting in successional replacement of different species in the same pond, or specific restriction to unlike habitats in other cases.

Broch (1969) found that the fairy shrimp *Branchinecta campestris* regulates hyperosmotically at environmental salinities of 286 mOsm/liter or less, but conforms from 451 to 660 mOsm/liter. This limits it to the early, more dilute phase in the seasonal cycle of the impermanent saline pond he studied. *Artemia salina,* a species that can regulate hypoosmotically (see Bayly, 1972, for review), utilizes the later, more concentrated period during the pond's terminal stages.

Of two anostracans occurring in temporary ponds around Lake Corangamite, Australia, Geddes (1973) found that *Branchinella australiensis* habitats ranged from 150 to 1200 ppm (4 to 34 mOsm) while those occupied by *Branchinella compacta* were 1530 to 15,900 ppm (43 to 460 mOsm). Physiological study of the two species revealed *B. australiensis* is a hyperosmotic regulator that dies in salinities approaching isosmocity (about 130 mOsm/kg of water), whereas *B. compacta* regulates hyperosmotically up to 160 mOsm/kg (its isosmotic point) and then becomes an osmoconformer up to 460 mOsm/kg. The hemolymph of *B. compacta* drops to 104 mOsm/kg in dilute medium (10 mOsm/kg), suggesting some loss of regulatory ability in low salinity environments.

Laboratory tests often show that adults, larvae, and eggs have contrasting tolerances. A good example shown by Belk (1972) concerns a conchostracan. In the laboratory, eggs hatch in saline waters that are hostile to survival of the larvae. The result is that this clam shrimp is limited to dilute waters, despite the euryhalinity of the embryonated eggs. Conversely, adults of certain other phyllopods can tolerate conditions that prevent egg hatching (Horne, 1967; Brown and Carpelan, 1971).

DISPERSAL POWERS

It has been stated that most temporary-pond forms have remarkable powers of dispersal (Bayly and Williams, 1973, p. 184). Passive dispersal is enhanced by the presence of drought-resistant stages, although it is not easy to evaluate the efficiency of any method. Desert dust storms distribute the resting stages of many species, and birds and flying insects serve as effective agents of transport (Maguire, 1963; Proctor et al., 1967).

Watson (1969) found all nine species of dragonflies that had naiad stages in temporary ponds in the northwest of Western Australia had unusual adult dispersal abilities. Contrasting with this, only 25 percent of the species from permanent waters could be so rated.

The generality, however, that occupants of impermanent aquatic habitats have noteworthy means of dispersal does not always hold true. Frey (1971) remarked on the variation shown among different populations of the cladoceran *Saycia,* emphasizing the isolated nature of temporary ponds. Restricted gene flow is implied by his conclusions. Belk (1974) found morphologic differences among conspecific populations of *Branchinecta packardi* in rock pools of northeastern Arizona, populations separated by no more than 22 km. The Australian conchostracan, *Limnadia stanleyana,* is often absent from some astatic pools in sandstone depressions that are very close to similar pools containing the species. Because of this, Bishop (1967b) was of the opinion that it has rather poor dispersal ability. Powers of dispersal are far from uniform throughout the many species that occupy temporary ponds. This could account for the occurrence of physiologic races and some discrepancies that appear in published accounts on the biology of inhabitants of temporary waters.

LITERATURE CITED

Bayly, I. A. E. 1964. A revision of the Australasian species of the freshwater genera *Boeckella* and *Hemiboeckella* (Copepoda: Calanoida). Aust. J. Mar. Freshwat. Res. 15:180–238.

———. 1972. Salinity tolerance and osmotic behavior of animals in athalassic saline and marine hypersaline waters. In R. F. Johnston (ed.), Annual review of ecology and systematics, Vol. 3, pp. 233–368. Annual Reviews Inc., Palo Alto, Calif.

———, and W. D. Williams. 1973. Inland waters and their ecology. Longman Australia. 316 p.

Belk, D. 1970. Functions of the conchostracan egg shell. Crustaceana 19:105–106.

———. 1972. The biology and ecology of *Eulimnadia antlei* Mackin (Conchostraca). Southwest. Nat. 16:297–305.

————. 1973. Suggestions of a timing mechanism inhibiting hatching after the first day of wetting in *Branchinecta lindahli* Packard eggs. Amer. Zool. 13:1339 (Abstr. 470).

————. 1974. Zoogeography of the Arizona Anostraca with a key to the North American species. Ph.D. Dissertation. Arizona State University, Tempe, Ariz. 90 p.

————, and M. S. Belk. In press. Hatching temperatures and new distributional records for *Caenestheriella setosa* (Crustacea, Conchostraca). Southwest. Nat.

Birge, E. A., and C. Juday. 1908. A summer resting stage in the development of *Cyclops bicuspidatus* Claus. Trans. Wisc. Acad. Sci., Arts and Letters 16:1–9.

Bishop, J. A. 1967a. Seasonal occurrence of a branchiopod crustacean, *Limnadia stanleyana* King (Conchostraca) in eastern Australia. J. Animal Ecol. 36:77–95.

————. 1967b. Some adaptations of *Limnadia stanleyana* King (Crustacea: Branchipoda: Conchostraca) to a temporary freshwater environment. J. Animal Ecol. 36:599–609.

————. 1968. Resistance of *Limnadia stanleyana* King (Branchiopoda, Conchostraca) to desiccation. Crustaceana 14:35–38.

Brewer, R. H. 1964. The phenology of *Diaptomus stagnalis* (Copepoda: Calanoida): the development and the hatching of the egg stage. Physiol. Zool. 37:1–20.

Broch, E. S. 1965. Mechanisms of adaptation of the fairy shrimp *Chirocephalopsis bundyi* Forbes to the temporary pond. Cornell Univ. Agr. Exp. Sta. Mem. 392:1–48.

————. 1969. The osmotic adaptations of the fairy shrimp *Branchinecta campestris* Lynch to saline astatic waters. Limnol. Oceanogr. 14:485–492.

Brown, L. R., and L. H. Carpelan. 1971. Egg hatching and life history of a fairy shrimp *Branchinecta mackini* Dexter (Crustacea: Anostraca) in a Mohave Desert playa (Rabbit Dry Lake). Ecology 52:41–54.

Carlisle, D. B. 1968. *Triops* (entomostraca) eggs killed only by boiling. Science 161:279.

Champeau, A. 1966. États de quiescence détermines chez les Copépodes d'eau saumâtre par les variations de chlorinité. Comptes Rendu Acad. Sci. Paris 262D:1289–1291.

————. 1967. États de quiescence détermines chez le Copépode Harpacticoide *Cletocamptus retrogressus* Schmankevetisch par des variations de chlorinité de température et par dessiccation. Comptes Rendu Acad. Sci. Paris 265D:248–251.

Clegg, J. S. 1964. The control of emergence and metabolism by external osmotic pressure and the role of free glycerol in developing cysts of *Artemia salina*. J. Exp. Biol. 41:879–892.

————. 1967. Metabolic studies of cryptobiosis in encysted embryos of *Artemia salina*. Comp. Biochem. Physiol. 20:801–809.

Cloudsley-Thompson, J. L. 1966. Orientation responses of *Triops granarius* (Lucas) (Branchiopoda: Notostraca) and *Streptocephalus* spp. (Branchiopoda: Anostraca). Hydrobiologia 27:33–38.

Cole, G. A. 1953. Notes on copepod encystment. Ecology 43:208–211.

———. 1966. Contrasts among calanoid copepods from permanent and temporary ponds in Arizona. Amer. Midl. Nat. 76:351–368.

———. 1968. Desert limnology. In G. W. Brown, Jr. (ed.), Desert biology, Vol. I pp. 423–486. Academic Press, New York.

Comita, G. W. 1972. The seasonal zooplankton cycles, production and transformation of energy in Severson Lake, Minnesota. Arch. Hydrobiol. 70:14–66.

Davis, C. C. 1967. Hatching processes in the eggs of aquatic invertebrates. Verh. Internat. Verein. Limnol. 16:1685–1689.

Decksbach, N. K. von. 1929. Zur Klassification der Gewässer vom astatischen Typus. Arch. Hydrobiol. 20:399–406.

Deevey, E. S. 1941. Notes on the encystment of the harpacticoid copepod *Canthocamptus staphylinoides* Pearse. Ecology 22:197–200.

Dussart, B. 1967. Les Copépodes des eaux continentales d'Europe occidentale I: Calanoïdes et Harpacticoïdes. Éditions B. Boubée & Cie, Paris. 500 p.

Edward, D. H. 1964. A cryptobiotic chironomid from south-western Australia Australian Soc. Limnol. Newsletter 3:29–30.

———. 1968. Chironomidae in temporary freshwaters. Australian Soc. Limnol Newsletter 6:3–5.

Elgmork, K. 1967. Ecological aspects of diapause in copepods. Proc. Symp Crustacea, Marine Biol. Assoc. India, Part 3:947–954.

Eriksen, C. H. 1966. Diurnal limnology of two highly turbid puddles. Verh. Internat. Verein. Limnol. 16:507–514.

Fairbridge, W. S. 1945. West Australian species of freshwater calanoids. I. Three new species of *Boeckella* with an account of the developmental stages of *B. opaqua,* n. sp. and a key to the genus. J. Proc. Roy. Soc. West. Australia 29:25–65.

Frey, D. G. 1971. Worldwide distribution and ecology of *Eurycercus* and *Saycia* (Cladocera). Limnol. Oceanogr. 16:254–308.

Fryer, G., and W. J. P. Smyly. 1954. Some remarks on the resting stages of some freshwater cyclopoid and harpacticoid copepods. Ann. Mag. Nat. Hist. (Ser 12) 7:65–72.

Gadgil, M., and O. T. Solbrig. 1972. The concept of *r*- and *K*-selection: Evidence from wild flowers and some theoretical considerations. Amer. Nat. 106:14–31.

Gauthier, H. 1928. Recherches sur la faune des eaux continentales de l'Algérie et de la Tunisie. Minerva, Algiers, Algeria.

Geddes, M. C. 1973. Salinity tolerance and osmotic and ionic regulation in *Branchinella australiensis* and B. compacta (Crustacea: Anostraca). Comp. Biochem. Physiol. 45A:559–569.

Gehrs, C. W., and B. D. Martin. 1974. Production of resting eggs by *Diaptomus clavipes* Schacht (Copepoda, Calanoida). Amer. Midl. Nat. 91:486–488.

Grainger, J. N. R. 1958. First stages in the adaptation of poikilotherms to temperature change. In C. L. Prosser (ed), Physiological adaptation, pp. 79–91. American Physiological Society, Washington, D.C.

Gurney, R. 1931. British freshwater Copepoda. Calanoida, I. The Ray Society, London. 238 p.

Hartland-Rowe, R. 1972. The limnology of temporary waters and the ecology of

Euphyllopoda. In R. B. Clark and R. J. Wootton (eds.), Essays in hydrobiology (presented to Leslie Harvey), pp. 15–31. University of Exeter Press, Exeter, England.

Hillyard, S. D., and A. Vinegar. 1972. Respiration and thermal tolerance of the phyllopod crustacea *Triops longicaudatus* and *Thamnocephalus platyurus* inhabiting desert ephemeral ponds. Physiol. Zool. 45:189–195.

Hinton, H. E. 1960. Cryptobiosis in the larva of *Polypedilum vanderplanki* Hint. (Chironomidae). J. Ins. Physiol. 5:286–300.

Horne, F. 1967. Effects of physical–chemical factors on the distribution and occurrence of some southeastern Wyoming phyllopods. Ecology 48:472–477.

———. 1968. Survival and ionic regulation of *Triops longicaudatus* in various salinities. Physiol. Zool. 41:180–186.

———. 1971. Some effects of temperature and oxygen concentration on phyllopod ecology. Ecology 52:343–347.

———, and K. W. Beyenbach. 1971. Physiological properties of hemoglobin in the branchiopod crustacean *Triops*. Amer. J. Physiol. 220:1875–1881.

Horsfall, W. R. 1956. Eggs of floodwater mosquitoes. III. Conditioning hatching of *Aëdes* mosquito eggs. Ann. Entomol. Soc. Amer. 49:66–77.

Hutchinson, G. E. 1967. A treatise on limnology, Vol. II. Introduction to lake biology and the limnoplankton. Wiley, New York. 1115 p.

Hynes, H. B. N. 1955. Biological notes on some East African aquatic Heteroptera. Proc. Roy. Entomol. Soc. London. A30:43–54.

Keilin, D. 1959. The problem of anabiosis or latent life: history and current concept. Proc. Roy. Soc. B150:149–191.

Lees, A. D. 1955. The physiology of diapause in arthropods. Cambridge University Press, New York, 151 p.

Maguire, B., Jr. 1963. The passive dispersal of small aquatic organisms and their colonization of isolated bodies of water. Ecol. Monogr. 33:161–185.

Mattox, N. T. 1950. Notes on the life history and description of a new species of conchostracan phyllopod, *Caenestheriella gynecia*. Trans. Amer. Microscop. Soc. 69:50–53.

Mayhew, W. W. 1968. Biology of desert amphibians and reptiles. In G. W. Brown, Jr. (ed.), Desert biology, Vol. I, pp. 195–356. Academic Press, New York.

Moore, W. G. 1955a. The life history of the spiny-tailed fairy shrimp in Louisiana. Ecology 36:176–184.

———. 1955b. Observations on heat death in the fairy shrimp, *Streptocephalus seali*. Proc. Louisiana Acad. Sci. 18:5–12.

———. 1967. Factors affecting egg-hatching in *Streptocephalus seali* (Branchiopoda, Anostraca). Proc. Symp. Crustacea, Marine Biol. Assoc. India, Part 2:724–735.

———. 1970. Limnological studies of temporary ponds in southeastern Louisiana. Southwest. Nat. 15:83–110.

———, and A. Burn. 1968. Lethal oxygen thresholds for certain temporary pond invertebrates and their application to field situations. Ecology 49:349–351.

O'Brien, F. I., J. M. Winner, and D. K. Krochak. 1973. Ecology of *Diaptomus leptopus* S. A. Forbes 1882 (Copepoda: Calanoida) under temporary pond conditions. Hydrobiologia 43:137–155.

Pancella, J. R., and R. G. Stross. 1963. Light induced hatching of *Daphnia* resting eggs. Chesapeake Sci. 4:135–140.

Peters, N., Jr. 1963. Embryonale Anpassungen oviparer Zahnkarpfen aus periodisch austrocknenden Gewässern. Intern. Rev. Hydrobiol. 48:257–313.

Pianka, E. R. 1972. *r* and *K* selection or *b* and *d* selection? Amer. Nat. 106:581–588.

Proctor, V. W., C. R. Malone, and V. L. DeVlaming. 1967. Dispersal of aquatic organisms: Viability of disseminules recovered from the intestinal tract of captive killdeer. Ecology 48:672–676.

Prophet, C. W. 1963. Physical–chemical characteristics of habitats and seasonal occurrence of some Anostraca in Oklahoma and Kansas. Ecology 44:798–801.

Røen, U. 1955. On the number of eggs in some free-living freshwater copepods. Verh. Internat. Verein. Limnol. 12:447–454.

Ruibal, R., L. Tevis, Jr., and V. Roig. 1969. The terrestrial ecology of the spadefoot toad *Scaphiopus hammondii*. Copeia 1969:571–584.

Rzóska, J. 1961. Observations on tropical rainpools and general remarks on temporary waters. Hydrobiologia 17:265–286.

Sawchyn, W. W., and U. T. Hammer. 1968. Growth and reproduction of some *Diaptomus* spp. in Saskatchewan ponds. Canadian J. Zool. 46:511–520.

Shan, R. K. 1969. Life cycle of a chydorid cladoceran, *Pleuroxus denticulatus* Birge. Hydrobiologia 34:513–523.

Sublette, J. E., and M. Sublette. 1967. The limnology of playa lakes on the Llano Estacado, New Mexico and Texas. Southwest. Nat. 12:369–406.

Sussman, A. S., and H. D. Halvorson. 1966. Spores: their dormancy and germination. Harper & Row, New York. 354 p.

Tribbey, B. A. 1965. A field and laboratory study of ecological succession in temporary ponds. Ph.D. Dissertation. University of Texas, Austin, Tex.

Watson, J. A. L. 1969. Taxonomy, ecology, and zoogeography of dragonflies (Odonata) from the north-west of Western Australia. Australian J. Zool. 17:65–112.

WATER AND ENERGY BUDGETS OF FREE-LIVING ANIMALS: MEASUREMENT USING ISOTOPICALLY LABELED WATER

Kenneth A. Nagy

*Laboratory of Nuclear Medicine and
Radiation Biology, and Department of Biology,
University of California, Los Angeles, California*

Abstract

Physiological measurements on free-living animals, until recently, have been possible only by various indirect estimates. However, techniques have now been developed to measure water, energy, and material fluxes in field animals using isotopically labeled water. Gross water fluxes can be measured with tritium- or deuterium-labeled water. In nondrinking animals, water is taken in as part of the food and is formed as a metabolic end product. If the rate of metabolic water production can be measured, dietary water input can be obtained by difference. Feeding rate can be calculated from dietary water input if the diet and its water content are known. Analysis of the field diet for water, caloric content, nitrogen, and salts provides a means for estimating the input rates of these substances in free-living animals. If the field animals being investigated are in a steady-state condition, output rates can also be obtained. Loss rates of these substances via specific avenues can be estimated from laboratory measurements. Metabolic rates of free-livng animals can be measured with doubly labeled water. Because the oxygen of body water is lost as water and CO_2, the difference between hydrogen isotope and oxygen isotope loss rates is a measure of the rate of CO_2 production. Equations to describe isotope fluxes are given, and improvements in isotope analysis are discussed.

The techniques outlined were used to investigate the relationship between seasonal changes in food quality and the physiology and behavior of a desert lizard, *Sauromalus obesus.* From spring to summer in 1970, the plant food of this lizard dehydrated, the electrolyte concentration increased, and nitrogen content decreased. Lizards began to lose weight in the middle of May, not because energy was limiting, but because the food was too dry. The drier the plants became, the more negative the water balance for these animals. In apparent response to this, the lizards ceased eating in June and July and "estivated" in rock crevices until October when they entered underground burrows for the winter. This behavior reduced estimated energy expenditure by about 75 percent, and reduced estimated water losses by about 90 percent. Despite the physiological and behavioral adaptations, these lizards showed a loss of about 40 percent of their body weight for the 1970 activity season.

INTRODUCTION

Environmental physiologists seek to understand the nature of the environment in which animals live, and the physiological and behavioral traits of animals which enhance their ability to survive in that environment. Because of difficulties involved in making appropriate measurements in the field, most previous studies have been done in the laboratory. Results of such studies include descriptions of physiological capacities in animals that have adaptive significance. Attempts to evaluate the role these capacities play in free-living animals have often been frustrating; however, recently developed techniques, such as radiotelemetry and isotopic tracers, have made possible the measurement of some physiological variables in noncaptive individuals (for recent surveys, see Mackay, 1970; Gessaman, 1973).

With the discovery by Lifson et al. (1949) that oxygen of expired CO_2 was in isotopic equilibrium with oxygen in the body water of rats and mice that had been enriched with $H_2^{18}O$, it became apparent that rates of CO_2 production, and hence metabolic rates, could be measured with isotopic water. Over the next several years Lifson and coworkers developed and tested methods for measuring energy and material balance in mammals using deuterium- and ^{18}O-labeled water (see Lifson and McClintock, 1966, for a review and references). In this paper, I shall consider additional procedures for accomplishing these measurements and discuss results of one such study of a free-living desert reptile. The methods and results summarized here are discussed in more detail elsewhere (Nagy, 1972; Nagy and Shoemaker, 1975).

METHODOLOGY

Body water molecules are directly involved in two important physiological processes: energetics and water relations. If both the oxygen and hydrogen atoms of body water are labeled with isotopes, water and energy kinetics can be measured. The hydrogen isotope is lost from the animal primarily in the form of water; the oxygen isotope is lost primarily as water and CO_2. Thus, measurement of the decrease in specific activity of these isotopes with time provides an assessment of the rates of water flux and energy metabolism. Moreover, in some circumstances it is possible to partition water input and thereby obtain an estimate of dietary water input, and, hence, feeding rate.

Budgets

Terrestrial vertebrates gain water by eating, drinking, and from oxidation of foodstuffs. Water is lost by evaporation, in the feces and urine, and in glandular secretions of some animals. Rates of gross water input and gross water output can be measured with deuterium- or tritium-labeled water in free-living animals. For many animals, particularly those living in deserts, free water is unobtainable, and water input consists entirely of dietary water and oxidation water. If the rate of metabolic water production can be estimated, the rate of dietary water input can be obtained by difference.

Currently, there are three ways to estimate the rate of metabolic water production. Animals can be prevented from eating in the field by fitting them with muzzles or by some other manipulation. In these animals, gross water input, measured by isotope kinetics, equals metabolic water production. This method may produce atypical results because muzzled animals may well have altered metabolic rates. Also, because the rate of water input in muzzled animals is low, the specific activity of the isotope decreases very slowly, and detectable differences between initial and final activities in body water may be hard to measure accurately. This can be alleviated somewhat by increasing the duration of the measurement period, but many animals would be adversely affected as a result.

Metabolic water production can be estimated from total water flux in nondrinking animals if they are in a steady-state condition, because metabolic water will be a constant fraction of the total flux: $m1 H_2O_M/day = k \times m1\ H_2O_T/day$, as long as the water content of the diet is constant. The fraction k represents the ratio of metabolic water produced per gram of food consumed to the total water gained per gram of food (preformed plus metabolic water). Thus,

$$\text{ml } H_2O_M/\text{day} = \frac{\text{ml } H_2O_M/\text{g food}}{\text{ml } H_2O_M/\text{g food} + \text{ml } H_2O_P/\text{g food}} \cdot \text{ml } H_2O_T/\text{day} \quad (1)$$

The amount of metabolic water produced per gram of food can be determined from laboratory measurements of assimilation (kcal assimilated/g food) and standard values for oxidation water yield of various foodstuffs (ml H_2O_M/kcal) with the relation of ml H_2O_M/g food = (kcal assimilated/g food) × (ml H_2O_M/kcal assimilated).

Another way of estimating metabolic water production is by measuring the rate of CO_2 production in field animals with doubly labeled water, and estimating water production from these values using standard conversion factors. Thus, it is possible to obtain an itemized water budget on the input side and a gross water output rate for free-living animals that do not drink water. For drinking animals, similar measurements are possible if the animals are released and recaptured between drinking bouts. If a drinking animal is sampled shortly before and after it drinks, the amount of water drunk can also be estimated from isotope kinetics. To partition the efflux side of the water budget, animals can be fed natural diets in the laboratory, where excretory and evaporative water losses can be measured. These results can be applied to the field animals with appropriate assumptions.

Material budgets can be obtained from feeding rates of field animals and laboratory determinations of food utilization. Food ingestion rates in the field are calculated from the water budget and composition of the diet. If the diet is known, its water content can be measured. If the rate of dietary water input is known, the rate of dry matter input (feeding rate) can be estimated with the equation

$$g \text{ dry food/day} = \frac{\text{ml } H_2O_P \text{ input/day}}{\text{ml } H_2O_P/g \text{ dry food}} \tag{2}$$

The rates of dry matter outputs in the form of urine and feces in field animals can be calculated from laboratory budgets by assuming that dry matter assimilation is the same in laboratory and field animals utilizing the same dietary components.

Food ingestion rates can also be estimated a second way. If metabolic rates in field animals are known from doubly labeled water measurements, and if field animals are in steady state with respect to energy, the rate of food input required to support the observed metabolic rate can be computed. This technique also requires feeding experiments and knowledge of the field diet, so that the amount of metabolizable energy available in the food per unit dry weight can be established.

Energy budgets can be calculated from water flux data or from doubly labeled water measurements, when they are coupled with feeding experiments. Both methods provide a means for calculating feeding rate, and gross energy input is the product of feeding rate and energy content of the food. The fractions of ingested energy that are excreted as feces and as urine are known from the feeding trials and can be converted to rates of energy output by multiplying by field feeding rate. Metabolic rate can be calculated as the difference (in hydrogen isotope studies), or it can be measured with doubly labeled water.

Once the rate of feeding and the diet composition are known, the input rates of nitrogen, salts, minerals, or any other material can be obtained, and output rates for these substances can be estimated from laboratory data.

Water Flux Calculations

When an animal loses water, the specific activity of the isotope in body fluids does not change as a result of water loss alone, because the activity of the exiting water is the same as that in the body. The specific activity of the body fluids does decrease, however, as a result of water gain, because the incoming water is not labeled. Because the absolute amount of isotope lost per unit time is a function of the specific activity of the body fluid at that time, the specific activity of body fluids decreases exponentially with time. This curve is described by the equation

$$k = \frac{\ln C_1 - \ln C_2}{t} \tag{3}$$

where C_1 and C_2 are initial and final specific activities, t is time elapsed, and k is a rate constant, which is called the fractional turnover rate. If a hydrogen isotope is used, k can have the units ml H_2O turned over/(ml body water $\times$ day), or any other unit of water amount or time unit desired, as long as these are used consistently. In the event that the animal is in a steady-state condition with respect to water (input = output), k is a measure of input and output rates, because input = turnover = output. If the animal is either storing water or dehydrating, input $\neq$ output, and k does not equal either but lies somewhere in between.

Lifson and McClintock (1966) examined this problem and derived an equation to describe water efflux when an animal's body water volume is changing linearly with time:

$$r_{H_2O} = \frac{\Delta N}{t} \frac{k_{2H^\cdot} - k_N}{k_N} \tag{4}$$

where r_{H_2O} is the rate of water loss in mmol/mmol body water per unit time, ΔN is the change in body water volume in mmol, $k_{2H^\cdot} = -(\Delta \ln C)/t$, where C is the specific activity of hydrogen isotope and t is time elapsed, and $k_N = (\Delta \ln N)/t$. After expanding this equation and including appropriate conversion factors to yield results in terms of ml H_2O lost/kg $\times$ day, the equation takes the form

$$\frac{\text{ml lost}}{\text{kg} \times \text{day}} = \frac{W_2 - W_1}{t} \frac{\ln(C_1/C_2) - \ln(W_2/W_1)}{\ln(W_2/W_1)} \frac{2000}{BW_1 + BW_2} \tag{5}$$

where W_1 and W_2 are initial and final body water volumes in milliliters, and BW_1 and BW_2 are initial and final body weights in grams. Water input rate can be calculated with the equation

$$\frac{\text{ml gain}}{\text{kg} \times \text{day}} = \frac{\text{ml lost}}{\text{kg} \times \text{day}} + \left(\frac{W_2 - W_1}{t} \cdot \frac{2000}{BW_1 + BW_2}\right) \tag{6}$$

Metabolic Rate Calculations

In an animal containing doubly labeled water, the oxygen isotope is lost primarily as water and in CO_2; the hydrogen isotope leaves the body primarily as water. Therefore, in a steady-state animal, the difference between the fractional turnover rates of oxygen and hydrogen isotopes is a measure of the rate of CO_2 production. Lifson and McClintock (1966) give the following equation to describe this relationship:

$$r_{CO_2} = \frac{N}{2}(k_{O^*} - k_{2H^*})$$

where r_{CO_2} is the rate of CO_2 production in mmol per unit time, $k_{O^*} = -(\Delta \ln O^*)/t$, where O^* is the specific activity of the oxygen isotope (atom percentage excess if ^{18}O is used), and N is body water amount in mmol. These authors give the following equation to describe the situation in animals with linearly changing water volumes:

$$r_{CO_2} = \frac{\Delta N}{2t} \cdot \frac{k_{O^*} - k_{2H^*}}{k_N} \tag{7}$$

After expansion and inclusion of appropriate conversion factors, this equation becomes

$$\frac{\text{ml } CO_2 \text{ lost}}{\text{g h}} = \frac{W_2 - W_1}{t(BW_1 + BW_2)} \cdot \frac{\ln(O_1/O_2) - \ln(C_1/C_2)}{\ln W_2/W_1} \times 51.86 \tag{8}$$

where O_1 and O_2 are initial and final activities of oxygen isotope (atom percentage excess).

Potential Errors

The preceding equations include several assumptions about the behavior of isotopes, and calculated rates can be as much as 20 to 100 percent different from actual rates when these assumptions do not hold. Lifson and McClintock (1966) and Mullen (1973) give detailed discussions of these assumptions. The most important ones appear to be that (1) rates of influx and efflux are constant, (2) isotopically labeled

H_2O and CO_2 do not fractionate upon leaving the body, and (3) once a labeled molecule leaves the animal, it does not reenter the body. It is difficult to imagine a situation where a terrestrial vertebrate would have constant rates of water gain and loss, and large errors can result from this assumption. There is disagreement regarding the importance of isotope fractionation effects. Lifson and McClintock (1966) present modified forms of the equations, which include corrections for fractionation of $D_2^{18}O$ based on published differences in the equilibrium vapor pressures of H_2O, HDO, and $C^{18}O_2$. However, Pinson (1952) was unable to detect differences in the activity of tritium between blood water and pulmonary or cutaneous water vapor in humans labeled with HTO. The magnitude of the errors involved in these assumptions is currently being investigated.

Isotope Assay

Two isotopes of hydrogen can be used to label body water: deuterium and tritium. Deuterium is a stable isotope, and materials containing this substance are usually measured with a mass spectrometer. This may require involved sample preparation before measurement: vacuum distillation of body fluids to obtain pure water, and conversion of the hydrogen in the water to hydrogen gas. Then the ratio of molecules of mass 3 ($^1H\,^2H$) to molecules of mass 2 ($^1H_2$) is measured via mass spectrometry, correcting for ^3H formed at the source. Deuterium-labeled water is readily available and is inexpensive, but deuterium analysis can be costly. Tritium-labeled water is also easily available and inexpensive. Because tritium is radioactive (soft beta particle emitter; half-life = 12.3 years) it is easily measured with a liquid scintillation counter. Body fluids can be distilled or counted directly, so sample preparation is simple. However, because tritium is radioactive, legal problems can arise in connection with the use of this isotope in the field. In these cases, deuterium may be the isotope of choice.

There are many radioactive isotopes of oxygen, but the half-lives are all very short (seconds). Two stable isotopes exist: ^{17}O with a mean natural abundance in water of 0.037 atom percent and ^{18}O (0.204 atom percent in naturally occurring water). Both isotopes are available in the form of enriched water, and both are relatively expensive. The ^{18}O content of body fluids can be measured by mass spectrometry. Again, this requires distillation of body fluids, and then conversion of the water to CO_2 with guanidine hydrochloride (Boyer et al., 1961; Mullen, 1973). The mass ratio 46 : 44 ($C^{16}O^{18}O : C^{16}O^{16}O$) is then measured, and the atom percentage excess ^{18}O calculated.

Alternatively, ^{18}O content of water samples can be determined by proton activation of ^{18}O to ^{18}F (Wood et al., 1975). The ^{18}F is a gamma emitter and can be counted with a NaI crystal gamma counter. With this analysis method, good results can be obtained from animals whose body water is enriched by only 0.1 atom percent ^{18}O excess.

RESULTS OF LABELED WATER STUDIES

Water and energy budgets are of great value in understanding the biology of wild animals for three primary reasons. First, by coupling budget information with measurements of the population density of the animal, the impact of that species on its environment can be determined and compared with the productivity of the habitat. Second, budgets for steady-state animals can yield information about minimum requirements for the species. This can be compared to water and energy fluxes measured at various times of the year to assess the effects of season on the welfare of these animals. Third, itemized budgets provide insights into physiological processes as they operate in the field, and permit examination of specific organ system functions in free-living animals in some detail.

Although many measurements of water flux have been made in free-living and captive animals, water budgets have been obtained in only a few instances. Laboratory water budgets have been determined for three marsupials (Hulbert and Dawson, 1974) and for calves (Phillips et al., 1971). Itemized budgets in animals living in outdoor enclosures are known for harbor seals (Depocas et al., 1971) and sheep (Brown and Lynch, 1972). In free-living vertebrates, water budgets have been obtained for two species of desert lizards (Minnich and Shoemaker, 1970; Nagy, 1972). Similarly, metabolic rates of free-ranging animals, as measured with doubly labeled water, are known only for pigeons (LeFebvre, 1964), purple martins (Utter and LeFebvre, 1970, 1973), four desert rodent species (Mullen, 1970, 1971a, b), and a lizard (Nagy and Shoemaker, 1975).

In the remainder of this section, I shall consider the energy and material budgets of a desert lizard (*Sauromalus obesus*) from the viewpoint of seasonal changes in desert resources and their relation to the dietary requirements of this animal. The results summarized here are presented in more detail elsewhere (Nagy, 1972, 1973; Nagy and Shoemaker, 1975).

Chuckwallas (*Sauromalus obesus*) are strictly herbivorous; their annual activity period extends from early spring to fall. In the central Mojave Desert, summers are quite hot and there is little or no rain. The condition of desert vegetation in spring and summer depends greatly on rainfall during the previous winter, and this varies extensively from year to year.

Desert Resources

Annual plants were fairly abundant in the central Mojave Desert in spring of 1970, despite a lower than average rainfall the previous winter. Perennial plants were lush. However, in May, maximum daily air temperatures increased rapidly, vegetation became dehydrated (Fig. 1), and most annual plants died. The water content of perennial plants decreased by more than two thirds, and their salt concentrations increased (Fig. 2). The nitrogen content of perennial plant leaves decreased in parallel with the decrease in succulence in late spring and summer (Fig. 3), but the weight specific energy content of plants did not change.

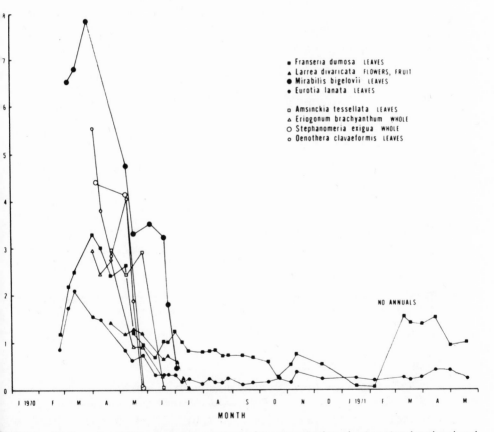

■ Franseria dumosa LEAVES
▲ Larrea divaricata FLOWERS, FRUIT
● Mirabilis bigelovii LEAVES
● Eurotia lanata LEAVES

□ Amsinckia tessellata LEAVES
△ Eriogonum brachyanthum WHOLE
○ Stephanomeria exigua WHOLE
○ Oenothera clavaeformis LEAVES

NO ANNUALS

MONTH

FIGURE 1 Seasonal variation in water content of representative plants eaten by chuckwal- Closed symbols represent perennials; open symbols indicate annual plants. (From gy, 1973.)

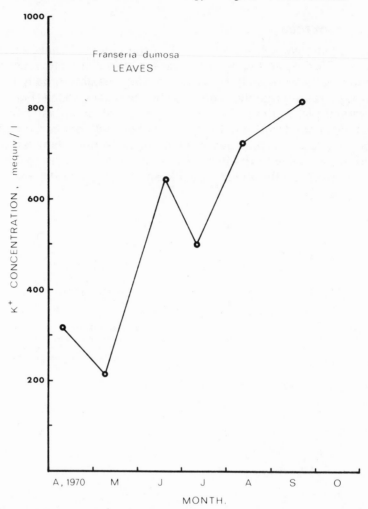

FIGURE 2   Seasonal variation in potassium concentration in leaves of an important dietary plant *Franseria* (= *Ambrosia*) *dumosa.*

Responses of Animals

In April and early May, chuckwallas were abroad for about 8 hours each day (Fig. 4). They spent only a few minutes feeding in morning and afternoon, but this was sufficient time to fill their stomachs. In June, total time spent abroad was much reduced, and only about a third of the animals present were abroad on any one day. The amount of time spent feeding was longer, presumably because green vegetation was harder

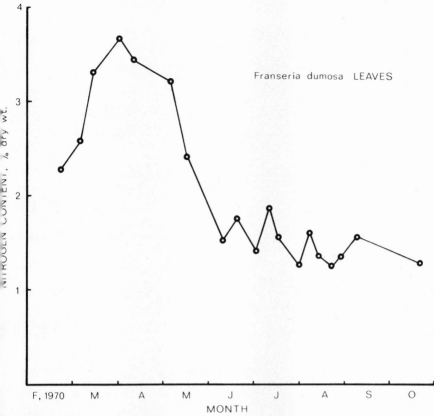

FIGURE 3 Seasonal variation in nitrogen content in leaves of *Franseria* (= *Ambrosia*) *dumosa*. (From Nagy and Shoemaker, 1975.)

to find. After early July, no lizards were seen feeding, and activity was restricted to basking for an hour or so at sunset. Although many animals were seen in rock crevices in October, none went abroad. After October, chuckwallas apparently entered underground burrows for the winter.

The rate of water flux in these lizards (indicated by the steepness of slopes in Fig. 5) was high in April. About the middle of May, water fluxes decreased because chuckwallas were eating drier food. However, these lizards were still ingesting energy at about the same rate as before. As summer progressed, the animals stopped feeding and reduced their activity (similar to estivation in some endotherms), and water fluxes declined even more. In summer, metabolic rates and thus energy requirements were reduced by about 75 percent as a result of this behavior (Fig. 6).

The net effect of seasonal changes in food quality and the animals'

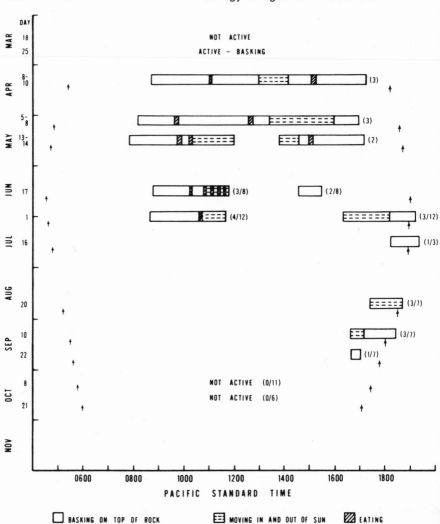

FIGURE 4   Daily behavior patterns of chuckwallas during their 1970 activity period. Arrows indicate sunrise and sunset. Numbers in parentheses for April and May indicate the number of lizards whose behavior was recorded. Thereafter, the fraction in parentheses indicates the number of lizards active over the number known to be present during presunrise censuses. Times are approximate. (From Nagy, 1973.)

responses can be assessed by examining changes in body weight (Fig 7). Chuckwallas gained weight in April, but they began losing weight in the middle of May. This decline continued into October, and animals entered winter burrows weighing only about 65 percent of their weight in early spring.

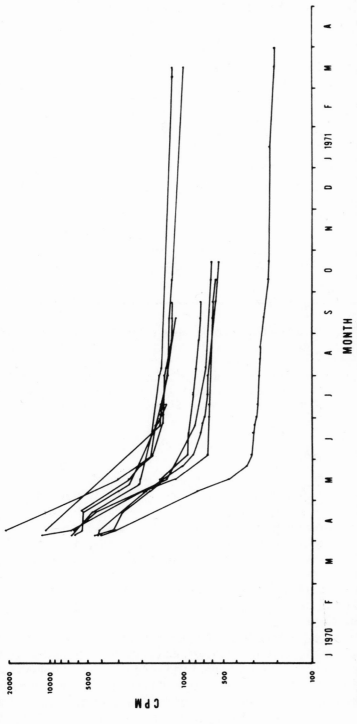

FIGURE 5 Seasonal changes in tritium disappearance in plasma of free-living chuckwallas. Counts per minute (cpm) values are from 10-μ1 plasma samples, and have been corrected for radioactive decay and adjusted to account for curve shifts due to reinjection of H³HO. (From Nagy, 1972.)

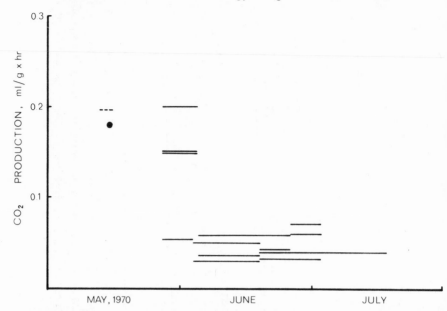

FIGURE 6 Rates of carbon dioxide production in free-living chuckwallas during spring and summer. Dashed line indicates CO_2 production calculated from water influx rates in lizards muzzled for 3 days in the field; filled circle is CO_2 production calculated from the energy budget (Fig. 8); solid lines represent rates of CO_2 production measured with doubly labeled water. Lines extend over the duration of each measurement period. (From Nagy and Shoemaker, 1975.)

Species Requirements

To determine energy and material requirements of an animal, it is essential to know fluxes in animals that are in a steady-state condition with respect to the substance of interest. In free-living animals, it is not possible to actually measure material input and output to determine if the animal is really in a steady-state condition. The most practical way of estimating the status of field animals is by measuring weight changes and assuming that the animal is near steady state when its weight remains constant. This estimate is probably fairly reliable with respect to water relations, because an animal's weight will likely change fastest as a result of water imbalance.

Determination of minimum requirements is more complicated, because even if the animal is in a steady-state condition, it could be taking in more than its minimum requirement and simply excreting the excess. In the case of water, it is relatively easy to determine minimum requirements, because excess water is excreted as fluid urine. Thus, when water loss is at its minimum in a steady-state vertebrate, the rate of

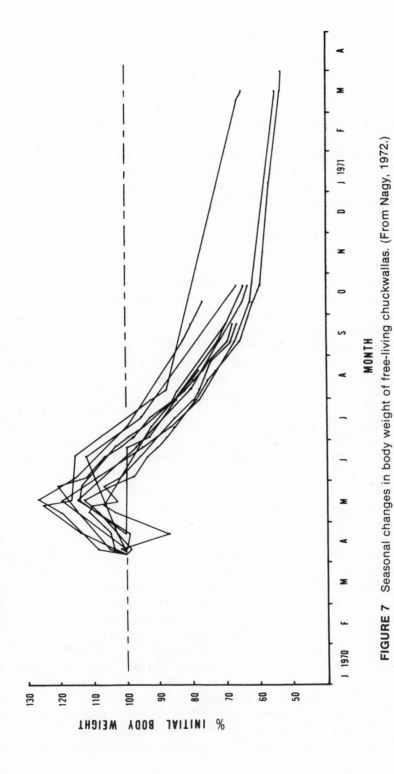

FIGURE 7 Seasonal changes in body weight of free-living chuckwallas. (From Nagy, 1972.)

water influx can be considered the minimum requirement for that animal. Excess energy is usually stored and is reflected as an increase in weight, so the rate of energy input in an animal with constant weight is probably close to the minimum energy requirement. To determine nitrogen and salt requirements, animals would have to be tested in the laboratory.

The lizards in this study had continually changing body weight (Fig. 7). However, it was possible to estimate the water fluxes in steady-state animals with data from early May (gaining weight) and late May (losing weight) by plotting water influx rate versus rate of body weight change (see Nagy, 1972). Similarly, the diet of a steady-state animal was calculated from May and June diets, and budgets were calculated from laboratory feeding experiments. These budgets are shown in Figure 8. As mentioned previously, the information in these budgets can be applied in a variety of ways. For present purposes, however, I shall use these results to evaluate the effects of changing food quality on the welfare of chuckwallas. More specifically, why did chuckwallas begin losing weight in the middle of May?

It can be suggested that chuckwallas lost weight because they were not obtaining sufficient energy. The energy budget indicates that a lizard requires 5.12 kcal/100 g × day, and this can be obtained by ingesting 1.24 g dry matter/100 g × day. Chuckwallas were eating as much food in late May as in April, and the energy content of the food did not change with season. Moreover, the fat content of these animals remained constant or increased slightly into early summer (Nagy, 1972), indicating that chuckwallas were in positive energy balance during that period. Thus, a shortage of energy was probably not responsible for weight losses in May. Nitrogen could have been in short supply, because the nitrogen content of food plants dropped markedly in May (Fig. 3). Unfortunately, it is not possible to determine nitrogen requirements from available data, so this question cannot be examined properly. The increasing salt concentration of the diet could have been forcing lizards to excrete more urine in order to osmoregulate. However, chuckwallas possess nasal salt glands that excrete about half the dietary potassium and almost all the chloride with very little water (Fig. 8). The other half of the dietary potassium was excreted in a precipitated form with urate, again with little accompanying water. Thus, chuckwallas excreted a considerable amount of electrolytes with little water loss, suggesting that increasing dietary salt concentrations did not induce weight loss.

Chuckwallas probably began losing weight in May because their diet became too dry to provide sufficient water for maintenance of water balance. As fluid urine volume was very low, the water budget in Figure 8 probably represents minimum water requirements. As the food plants

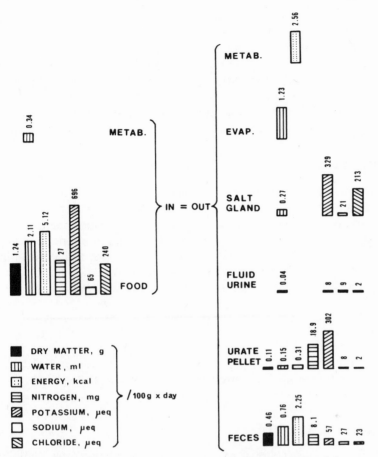

FIGURE 8 Energy and material budgets of chuckwallas maintaining constant body weights in the field. Steady-state conditions are assumed. (Modified from Nagy, 1972, and Nagy and Shoemaker, 1975.)

dehydrated, water input via the diet decreased, but water losses stayed the same. Dry matter intake stayed the same or slightly increased, and because chuckwallas cannot digest cellulose, about 40 percent of ingested dry matter was voided as feces. The feces was always about 60 percent water. As the diet became progressively drier, more and more water had to be taken from the body water pool to satisfy water loss requirements. Thus, chuckwallas lost weight, even though they were in positive energy balance. Because a great deal of water is lost in feces and by high evaporation associated with activity, a chuckwalla that is abroad and feeding has a much higher rate of water loss than an inactive, unfed animal. In fact, when the diet is very dry, the more a chuckwalla eats, the faster it dehydrates. In the field, chuckwallas responded to the decreasing water content of food plants by decreasing and eventually not eating, and by "estivating" in rock crevices. This behavior reduced their water losses by about 90 percent (Nagy, 1972), and energy expenditure was reduced about 75 percent (Fig. 6). Yet, in spite of all the physiological and behavioral adaptations these lizards have, they still showed a net loss of about 40 percent of their body weight in 1970. Certainly, the quality of the diet, which itself is dependent upon rainfall, is very important to the welfare of these lizards.

CONCLUSIONS

Field studies involving isotopically labeled water are feasible and profitable. When coupled with laboratory feeding experiments, field data can be used to estimate energy and material budgets for free-living animals. A year-long study can reveal the effects of changes in environmental resources on an animal, as well as the responses of the animal to these changes. The results of such a study are applicable in several ways, and are of immediate relevance in understanding how animals survive in nature.

Acknowledgments

Preparation of this paper was supported by Contract AT(04-1) GEN-12 between the University of California and the U.S. Atomic Energy Commission.

LITERATURE CITED

Boyer, P. D., D. J. Graves, C. H. Suelter, and M. E. Dempsey. 1961. Simple procedure for conversion of oxygen of orthophosphate or water to carbon dioxide for oxygen-18 determination. Anal. Chem. 33:1906–1909.

Brown, G. D., and J. J. Lynch. 1972. Some aspects of the water balance of sheep at pasture when deprived of drinking water. Aust. J. Agr. Res. 23:669–684.

Depocas, F., J. S. Hart, and H. D. Fisher. 1971. Sea water drinking and water flux in starved and fed harbor seals, *Phoca vitulina*. Can. J. Physiol. Pharmacol. 49:53–62.

Gessaman, J. A. (ed.). 1973. Ecological energetics of homeotherms; a view compatible with ecological modeling. Utah State Univ., Monograph Series, Vol. 20, 155 pp.

Hulbert, A. J., and T. J. Dawson. 1974. Water metabolism in perameloid marsupials from different environments. Comp. Biochem. Physiol. 47A:617–633.

LeFebvre, E. A. 1964. The use of D_2O^{18} for measuring energy metabolism in *Columba livia* at rest and in flight. Auk 81:403–416.

Lifson, N., G. B. Gordon, M. B. Visscher, and A. O. Nier. 1949. The fate of utilized molecular oxygen of respiratory carbon dioxide, studied with the aid of heavy oxygen. J. Biol. Chem. 180:803–811.

————, and R. McClintock. 1966. Theory of use of the turnover rates of body water for measuring energy and material balance. J. Theoret. Biol. 12:46–74.

Mackay, R. S. 1970. Bio-medical telemetry; sensing and transmitting biological information from animals and man, 2nd ed. Wiley, New York. 533 pp.

Minnich, J. E., and V. H. Shoemaker. 1970. Diet, behavior and water turnover in the desert iguana, *Dipsosaurus dorsalis*. Amer. Midl. Natur. 84:496–509.

Mullen, R. K. 1970. Respiratory metabolism and body water turnover rates of *Perognathus formosus* in its natural environment. Comp. Biochem. Physiol. 32:259–265.

————. 1971a. Energy metabolism and body water turnover rates of two species of free-living kangaroo rats, *Dipodomys merriami* and *Dipodomys microps*. Comp. Biochem. Physiol. 39A:379–390.

————. 1971b. Energy metabolism of *Peromyscus crinitus* in its natural environment. J. Mammal. 52:633–635.

————. 1973. The $D_2{}^{18}O$ method of measuring the energy metabolism of free-living animals. In Gessaman (1973), *op. cit.*, pp. 32–43.

Nagy, K. A. 1972. Water and electrolyte budgets of a free-living desert lizard, *Sauromalus obesus*. J. Comp. Physiol. 79:39–62.

————. 1973. Behavior, diet and reproduction in a desert lizard, *Sauromalus obesus*. Copeia 1973:93–102.

————, and V. H. Shoemaker. 1975. Energy and nitrogen budgets of a free-living desert lizard, *Sauromalus obesus*. Physiol. Zoo. (in press).

Phillips, R. W., L. D. Lewis, and K. L. Knox. 1971. Alterations in body water turnover and distribution in neonatal calves with acute diarrhea. Ann. N.Y. Acad. Sci. 176:231–243.

Pinson, E. A. 1952. Water exchanges and barriers as studied by the use of hydrogen isotopes. Physiol. Rev. 32:123–134.

Utter, J. M., and E. A. LeFebvre. 1970. Energy expenditure from free flight by the Purple Martin (*Progne subis*). Comp. Biochem. Physiol. 35:713–719.

————, and E. A. LeFebvre. 1973. Daily energy expenditure of Purple Martins (*Progne subis*) during the breeding season: estimates using D_2O^{18} and time budget methods. Ecology 54:597–604.

Wood, R. A., K. A. Nagy, N. S. MacDonald, S. T. Wakakuwa, R. J. Beckman, and H. Kaaz. 1975. Determination of oxygen-18 in water contained in biological samples by charged particle activation. Anal. Chem. 47:646–650.

LIMITATIONS IMPOSED BY DESERT HEAT ON MAN'S PERFORMANCE

D. B. Dill

Laboratory of Environmental Patho-Physiology, Desert Research Institute, University of Nevada System, Boulder City, Nevada

Abstract

Studies of workmen and members of our party in the desert in 1932 revealed that sweat volume can reach 10 liters/day and that heat cramps may occur in workmen especially during the first days on the job. Heat cramps arise from excessive loss of salt in sweat; treatment involves replacement of sodium chloride; potassium is not involved. Salt concentration in sweat varies widely in both sexes; it is not sex dependent. Those whose sweat is unusually salty undergo depletion of both water and salt in long walks even when satisfying their thirst, which suggests that osmotic pressure is the key to regulation of thirst.

The saltiness of sweat not only varies from one person to another but increases with rate of sweating and decreases with acclimatization. It is not closely dependent on skin or rectal temperature. The rate of sweating in desert walks is adjusted to the need for temperature regulation and maintained at that rate as long as conditions are unchanged. It is adequate even in men of 60 years but may be inadequate in women when heat stress is great and metabolic rate high. The skin and body temperatures of women increase more rapidly than men's, which suggests that woman's limiting value for sweat rate is lower than man's.

Men can walk in desert heat at 100 m/min for 1 hour with little discomfort, even without drinking, but a second hour without water results in increasing heart rate and rectal temperature and a dry mouth. If water and salt are replenished periodically, the second hour becomes easy. It is possible to drink and absorb more than 1 liter/hour during such a 2-hour walk.

INTRODUCTION

What happens to man during walks in heat? Men and women and boys and girls have walked hundreds of miles in desert heat to help find the answers. None has experienced any injury more unpleasant than sunburn or blistered feet. Specific questions raised concern the saltiness of sweat and factors that determine saltiness, the rate of sweating, the regulation of body temperature, and to what extent sex and age modify the answers. What occurs during acclimatization? What clue to successful performance is given by the heart rate? Finally, how do answers to these questions contribute to an understanding of deaths from exposure to desert heat?

Our studies of desert heat began at Boulder City in 1932, a year after severe effects of heat had been experienced by workmen on Hoover Dam. In 1931, the workmen lived in the canyon without air conditioning. Those on night shift particularly were unable to sleep comfortably. There were 13 reported deaths from heat breakdown. When our party from the Harvard Fatigue Laboratory arrived the next summer, both dormitories and the mess hall were air conditioned.

At that time physiologists had known for generations that heat cramps experienced by workmen in hot jobs resulted from excessive loss of salt in sweat. This fact had not become known to many industrial physicians, as was evident from the large sign facing diners in the mess hall: THE DOCTOR SAYS DRINK PLENTY OF WATER. Dr. Wales Haas, the contractor's surgeon, agreed to our suggestion of an addition: AND PUT PLENTY OF SALT ON YOUR FOOD.

During that summer of 1932 our physicians, Talbott and Michelsen, had a few cases of heat cramps to study at the contractor's hospital. They demonstrated beyond doubt that the cure was salt solution by mouth in most cases and intravenously or intradermally in severe cases. The salt supplied was pure sodium chloride: no other electrolytes were needed (Talbott and Michelsen, 1933).

We all exercised in the desert that summer, walking, climbing, or playing tennis. Some of us were on a constant diet, first at La Jolla, California, where sweating was minimal, and then at Boulder City (Dill et al., 1933b). The losses of electrolytes in sweat each 24-hour period were estimated from the difference between input in the diet and output in the urine as shown in Table 1. The results demonstrated the wide range in saltiness of sweat from one person to another. Also it was evident that those of us who exercised most and put out the saltiest sweat depleted their salt reserves and after an active day occasionally put out chloride- and sodium-free urine. The constant diet included about 210 meq of chloride, 198 of sodium, and 88 of potassium per 24 hours. Dill put out the saltiest sweat; over the first 6 days at Boulder City his urine electro-

TABLE 1 Daily water intakes and electrolyte outputs, control versus desert

	Water (liters)	Cl (meq)	Na (meq)	K (meq)
DBD				
Control[a]	1.75	210	198	89
Desert[b]	6.62	82	85	87
Δ[c]	4.87	128	113	2
BFJ				
Control	1.85	260	250	88
Desert	4.38	224	224	86
Δ	2.53	36	26	2
HTE				
Control	1.65	250	242	86
Desert	4.28	192	184	69
Δ	2.63	58	58	17
WC				
Control	1.75	208	195	83
Desert	5.08	136	117	78
Δ	3.33	72	78	5
FC				
Control	1.75	208	195	83
Desert	5.50	148	145	76
Δ	3.75	60	50	7

[a] Control observations on the constant diet were carried out in the cool climate of La Jolla, California, where activity and sweating were minimal. The mean 24-hour intake of water and outputs in the urine of chloride, sodium, and potassium were employed to estimate the effects of desert life on water and electrolyte exchanges.
[b] Mean 24-hour intake of water and outputs of chloride, sodium, and potassium in the urine during several days in the desert, July 1932.
[c] Mean 24-hour output in the sweat of water, chloride, sodium, and potassium during several days in the desert.

lytes expressed in meq per 24 hours ranged from 10 to 24 of chloride, 15 to 33 of sodium, and 56 to 97 of potassium. The mean deficits in output of those ions in urine, shown in Table 1, correspond to losses in sweat.

The mean extra water intake at Boulder City also is shown in Table 1 for each of five men. It is clear from the data of Table 1 that the daily extra water intake varied considerably from subject to subject, 4.87 liters in DBD to 2.53 in BFJ. These differences are approximate measures of daily outdoor activity and of sweat output. There were wide ranges in electrolyte deficits related to rate of sweating and saltiness of sweat. We concluded that at our level of activity major salt deficit is unlikely unless sweat is unusually salty as in the case of DBD. That was borne out by the fact that only five workmen during the summer of 1932 reported to the hospital with heat cramps. Even these five cases had unusual histories. Some were new to the job and unacclimatized. Others had been on a weekend binge with little to eat and little salt. That study by Talbott and Michelsen (1933) led to the conclusion that a criterion useful in diagnosis is rapid relief by parenteral injection of large amounts of 0.9 percent sodium chloride solution. Their paper is useful in reporting 26 earlier related studies including one by the pioneer in this field, Haldane, on "water poisoning" (Haldane, 1923).

During that summer we acquired a dog from the Las Vegas pound and compared his performance with mine in an all-day walk (Dill et al., 1933a). We followed a course at an easy pace of 3.5 mph that brought us back to the laboratory every 90 minutes. We were weighed and voluntary water intake was measured. This continued for four laps without food. I walked one more lap, but the dog's feet, despite Bull Durham tobacco bags, were painfully blistered so I left him behind. Much was learned from that venture into the field of comparative physiology. The dog regained his weight loss each time while I continued to lose. At the end of the fourth lap with our thirsts satisfied, I had lost 3.1 percent of my body weight and the dog had lost 100 g or 0.6 percent of his body weight. Since we ate nothing, his small loss was in part due to consumption of fat and carbohydrate. The interpretation of this difference is based on the hypothesis that one drinks enough to restore osmotic pressure. The dog distilled water from his lungs and respiratory passages, lost virtually no salt, and hence regained his weight each lap by drinking as much as he had evaporated. In my case salt was being lost in the sweat and restoration of my osmotic pressure required less water than I had lost in sweat. At the end of the fifth lap my total water intake was 6.1 kg and my weight loss was 3.1 kg. I had not urinated, so my sweat volume in 7.1 hours was approximately 9.2 liters. If our hypothesis is correct, I had a deficit of 3.1 liters of physiological sodium chloride solution, 0.9 percent, which contained 28 g of salt or about 475 meq. This implies that my 9.2 liters of sweat contained 475/9.2 or about 50 meq of chloride/liter. This is close to the concentration of chloride in my sweat collected in rubber gloves during that summer in eight games of tennis, 59 meq/liter.

Five years later some of us returned to Boulder City together with E. F. Adolph of the University of Rochester and F. G. Hall and Ross McFarland with whom I had been associated in a high-altitude study 2 years before. That summer we used the washdown method to collect sweat from the whole body. Sweat volume was known from weight loss, and chemical analysis of the washings permitted calculation of electrolyte composition of sweat (Dill et al., 1938). The findings will be discussed in the light of findings at Boulder City in 1964 when several of us who had been studied in 1932 or 1937 returned for follow-up studies. In the report of that study (Dill et al., 1966) it was concluded that concentration of chloride in sweat tends to increase with sweat rate but is not closely related to skin and rectal temperatures. It generally decreases with acclimatization but an overriding factor is individual idiosyncrasy. There may be a genetic factor: my son and I had unusually salty sweat; Hall and his son had unusually dilute sweat. Thirty-one boys volunteered for the 1-hour walk at 100 m/min and the washdown. These findings will be compared with later observations on 30 girls.

A new field of exploration was opened in 1966 when women joined our investigative team and also served as subjects. Kay Burrus of Indiana University and Gale Gehlsen of Ball State University, both athletic and both in departments of physical education, took part in all our studies and persuaded 30 girls to undergo the washdown after a 1-hour walk (Dill et al., 1967). The concentration of chloride in the girls' sweat showed about the same variability and about the same mean chloride concentration as found in boys' sweat studied 2 years before. The rate of sweating in relation to body surface was independent of age and sex, but depended on metabolic rate and ambient conditions, particularly air temperature, solar radiation, and wind. The composition of sweat collected in a rubber glove was compared with that of sweat from the entire body surface. The mean values in meq/liter for chloride, sodium, and potassium are shown in Table 2. The low concentration of potassium in sweat seen in Table 2 and the large intake of potassium in normal diets revealed in Table 1 make it unlikely that the potassium in elixirs athletes are urged to drink has any virtue except sales appeal.

Is the rate of sweating maintained during long walks in the desert? In our experience it is maintained, but investigators of exercise in laboratory hot rooms seems reluctant to accept this conclusion. In hot rooms, particularly when humidity is high, the sweat rate certainly does decline. The term *fatigue of the sweat glands* was coined to describe this phenomenon; later it was described as *sweat suppression.* In exploring this phenomenon we recorded rate of sweating every 7-minute lap in a

TABLE 2 Composition of sweat in relation to sex and age (mEq/liter)

	Cl	Na	K
30 girls	16.1	25.1	4.8
31 boys	20.7	27.8	8.0
Male, 20 yr	26.1	34.5	5.2
Male, 33 yr	18.2	24.5	5.2
Male, 34 yr	22.3	28.0	4.7
Male, 73 yr	65.4	85.7	5.1

closed desert course. In 12 men and 2 women the mean rates of sweat loss rose from 100 g/lap to about 130 in the third lap and did not change in the next four laps. There was no evidence of sweat suppression. It is reasonable to conclude that sweat suppression does not occur during a 1-hour walk in the dry heat of the desert since all sweat evaporates and is used for cooling. If the rate of sweating were to decline or if the subject were unable to sweat enough to regulate body temperature, breakdown would follow.

Some women are as able as some men to walk for 1 hour on Boulder City's hottest days. An unusually fit and slender girl of 15 who walked 4 miles daily to and from school was matched with an 18-year-old boy. They walked for 1 hour together on 10 consecutive weekdays with the same rise in rectal temperature, 0.9°C. When their sweat rates per square meter of body surface were compared, they turned out to be identical, 414 ml per hour and per square meter of body surface (Dill, 1972).

The experience of four women who undertook a 1-hour walk in the afternoon of a hot day was in sharp contrast to that of the 15-year-old girl just mentioned. The four included Kay Burrus (KB) and Gale Gehlsen (GG), both age 30 and both fit, subject A, age 18 and active, and subject B, age 49. This walk occurred in 1967 before we acquired telemetering equipment for heart rate. The girls studied in 1966 had walked at 80 m/min; this was the first time women had undertaken a 1-hour walk at 100 m/min. The results are shown in Table 3. Two sets of observations are notable: high skin temperature, range 36.9 to 38.6, mean 37.9, and high body temperature, 38.8 to 40.2, mean 39.5. All but subject B completed the hour; she was persuaded to stop after 53 minutes. A few years later when seven men were completing 2-hour walks on hot afternoons, KB joined us determined to complete the 2 hours. She was persuaded to stop after 80 min with a rectal temperature of 40.1°C and a heart rate of 184. The results indicate that many women are unable to walk at 100 m/min for 1 hour on a hot desert day and that few women are able to walk at that rate for 2 hours.

TABLE 3 Four women walking at 100 m/min, July 27, 1967 (ambient temperature 40°C, black body temperature 50°, relative humidity 30 percent, wind speed 5 to 10 mph)

	Age (yr)	Height (cm)	Weight (kg)	Duration (min)	Sweat rate (ml/min·m²)	Sweat chloride (meq/liter)	T_{re} (°C)	T_{skin} (°C)
A	18	165	48.1	65	7.6	30.0	40.1	38.2
GG	30	163	49.5	61	9.0	34.4	38.8	36.9
KB	30	167	61.5	64	7.7	42.0	39.1	38.0
B	49	163	58.4	53	7.4	15.3	40.2	38.6

Walks at 100 m/min for 2 hours were undertaken by eight males to evaluate the effects of replenishing salt and water (Dill et al., 1973). Seven were able to complete four such walks, two walks without drinking and two with replenishment of salt and water. In preliminary studies the saltiness of each man's sweat was determined and enough of an appropriate salt solution was prepared and cooled prior to each walk to use in replenishing losses. At the end of each 7-min lap the subject was handed an Erlenmeyer flask containing the calculated volume of salt solution of proper concentration. All found the walk easier when salt and water losses were replenished. The heart rates and rectal temperature leveled off; evidently, the slow rise without replenishment depends on dehydration with decreased volume of body fluids. The subjects noted that the first portions of fluid seemed to be emptied slowly by the stomach, but soon this feeling disappeared. An eighth, relatively unfit young man had trouble, vomiting before completing the first walk with replenishment. It appears that up to a limit of aerobic metabolism or a limit of environmental stress the emptying capacity of the stomach is adequate and, as proved by our experience, exceeds 1 liter/hour. In an extraordinary record of soldiers' performance in the heat, water was drunk at rates exceeding 2 liters/hour for up to 4 hours (Eishna et al., 1945).

The last experiments to be discussed were carried out on men about 60 years of age. They varied in fitness but the majority were unusually fit members of an adult fitness program at San Diego State University. We were interested in how well such men coming from the mild climate of San Diego would adapt to desert heat. We also wished to determine how well men of that age can regulate body temperature in desert walks. For comparison we studied 10 residents of Southern Nevada 60 years of age or older.

The response of these men varied widely. The members of the fitness program performed exceptionally well. All could walk for 1 hour at 100 m/min and all but one at 120 m/min. Temperature regulation was not a problem. The 10 Nevadans were unaccustomed to planned exercises except for some who played golf or swam. Only one walked for 1 hour at 120 m/min. Of the other nine, six were able to walk for 1 hour at 100 m/min. Increase in heart rate had an inverse relation to the quality of the individual's exercise regimen. In the less fit Nevadans the heart rate increased as breakdown approached, giving us a warning signal. Even the less fit subjects had no difficulty regulating body temperature.

The ease with which fit men of 60 years can walk for 1 hour at 120 m/min contrasts with the difficulty with which fit women walk for 1 hour at 100 m/min, environmental stress being about the same. Three significant indexes to performance in desert walks are sweat rate per unit of body surface, mean skin temperature, and increment in rectal tem-

TABLE 4

	Four women	Five men
Rate of walking (m/min)	100	120
Mean sweat rate (ml/min·m²)	7.9	12.5
Mean skin temperature (°C)	37.9	35.7
Increment in rectal temperature (°C)	2.1	1.4

perature. To emphasize the difference in responses of the sexes in desert walks, these indexes for four women are compared in Table 4 with indexes for five men over 60 walking 20 m/min faster than the women. Other factors may be involved, but it seems likely that women have a smaller capacity for sweating than men. The sweating rates attained by the five men are as great as we have found in young men in desert walks. Clearly, the capacity for temperature regulation shows no sign of diminishing with age.

These observations on healthy men, women, boys, and girls resulted in no ill effects, but they gave a better understanding of how death in desert heat can occur. Healthy men can walk 7.5 miles in 2 hours without water on the hottest summer day, but most women would collapse in the attempt. Collapse would come even sooner in man if the cardiovascular system is unfit, if body fat content is excessive, or if feet or joints fail. A major handicap is lack of experience, which may engender fear; this in turn may lead to irrational behavior such as walking away from a water source. A compilation of case histories of deaths in the desert would furnish an impressive guide for visitors to the desert.

Acknowledgments

This study was supported by NSF Grant GB-35281 and National Institutes of Health Grant HD-05625 to the Desert Research Institute, University of Nevada System. Colleagues and student assistants have been invaluable both in the laboratory and as subjects.

LITERATURE CITED

Dill, D. B. 1972. Desert Sweat Rates. In S. Itoh, K. Ogata, K. Takagi, and H. Yoshimura (eds.), Advances in climatic physiology, pp. 134–143. Igaku Shoin, Hongo, Bunkyo-ku, Tokyo.

———, A. V. Bock, and H. T. Edwards. 1933a. Mechanisms for dissipating heat in man and dog. Amer. J. Physiol. 104:36–43.

———, F. G. Hall, and H. T. Edwards. 1938. Changes in composition of sweat during acclimatization to heat. Amer. J. Physiol. 123:412–419.

————, F. G. Hall, and W. van Beaumont. 1966. Sweat chloride concentration: sweat rate, metabolic rate, skin temperature, and age. J. Appl. Physiol. 21:99–106.

————, S. M. Horvath, W. van Beaumont, G. Gehlsen, and K. Burrus. 1967. Sweat electrolytes in desert walks. J. Appl. Physiol. 23:746–751.

————, B. F. Jones, H. T. Edwards, and S. A. Oberg. 1933b. Salt economy in extreme dry heat. J. Biol. Chem. 100:755–767.

————, M. K. Yousef, and J. D. Nelson. 1973. Responses of men and women to two-hour walks in desert heat. J. Appl. Physiol. 35:231–235.

Eishna, L. W., W. F. Ashe, W. B. Bean, and W. B. Smiley. 1945. The upper limits of environmental heat and humidity tolerated by acclimatized man working in hot environments. J. Ind. Hyg. Toxicol. 27:59–84.

Haldane, J. S. 1923. Water poisoning. Brit. Med. J. i. 986.

Talbott, J. H., and J. Michelsen. 1933. Heat cramps. A clinical and chemical study. J. Clin. Invest. 12:533–549.

DESERT EXPANSION AND THE ADAPTIVE PROBLEMS OF THE INHABITANTS

J. L. Cloudsley-Thompson

Department of Zoology, Birkbeck College,
University of London, England

Abstract

The present drought in the Sahel savanna regions on the southern fringe of the Sahara is by no means abnormal. It is part of a drying-up process that took place mainly between 7000 and 3500 years B.P. Subsequent impoverishment of the flora and fauna has been due almost entirely to human activities: bad agriculture, felling trees for fuel, and overgrazing by domestic stock. Under the impact of shifting cultivation and the use of fire, forest has deteriorated into savanna; overgrazing has then degraded much of this into desert, as indicated by historical evidence. Various solutions have been proposed to the problem of desert expansion, but individually these are either impracticable or unpromising. The best hope for the future lies in multiple land use and industrial development.

INTRODUCTION

At present the Sahel savanna zone south of the Sahara is gripped by the most devastating drought within living memory. The Niger and other rivers have reached their lowest levels since recordings were first made; Lake Chad, always variable, has shrunk dramatically. It is, therefore, not surprising that the disaster should be looked upon with serious concern. The desert's fringe is a fragile, unstable ecosystem, naturally subject to periods of drought, which would not necessarily bring famine had not the region for centuries been misused and intensively over-

exploited by its human population. This overexploitation, greatly inten‐
sified during the last few decades, is responsible for the present calam‐
ity. By 1970 the land was supporting 24 million people and about the
same number of animals—roughly one third more people and twice as
many animals as 40 years ago (Wade, 1974).

The development of the Sahel savanna in countries such as Niger and
Mali, especially since they achieved independence, has been accom‐
panied by an increase in the number of wells, pumped water, and mass
vaccination of the people and their domestic animals. In consequence
the cattle, sheep, and goats of the nomads have increased enormously,
for wealth is assessed in numbers rather than by the quality of stock.
Much of the recent economic development and population expansion
has taken place during a period of abnormally high rainfall. This was
above the long-period average in 27 of the 39 years between 1920 and
1950 (Winstanley, 1973). What has made the subsequent drought so
devastating, however, has been the increase in domestic animals and
consequent overgrazing during those green years, rather than just the
subsequent period of reduced rainfall, which is no more than normal in
semiarid regions bordering desert. To obtain an objective assessment
of the situation, a very much longer period of time should be considered
(Grove, 1974).

CLIMATIC FACTORS

During late Pleistocene, the Sahara experienced both pluvial and
interpluvial periods when the climate was successively wetter and drier
than it is today. At the end of the Pleistocene, the regions bordering its
southern edge were richly supplied with lakes and active rivers, from
Mauritania in the west to the plains south of Khartoum. Several deposits
left by these bodies of water have been dated; most of them are between
5000 and 8000 years old. Although the lakes near Khartoum apparently
dried up 7000 years ago, many others lasted until 3000 before the
present (B.P.) (Kendall, 1969). As the climate became warmer and drier,
montane forest trees were replaced by xerophyllous species, and about
4800 years ago desiccation had increased to such an extent that the
mediterranean macchia, which had previously covered the lowland re‐
gions, could survive only on the higher mountains. It was replaced by
Sahelian flora, of which *Acacia* is a prominent genus, as far north as the
Atlas mountains. In historical times, real desert conditions set in, and
the central Sahara became inhospital to plants and animals (Bakker,
1963). These conclusions, based on pollen analyses, are supported by
study of the *goz* dune systems of Kordofan in central Sudan (Warren,
1970). Butzer (1959), however, speaks of a subpluvial period lasting until
about 4350 B.P.

The presence of an arid zone in northern Africa is almost certainly the

irect result of fundamental features of atmospheric circulation. The
limate of the Sahara today is controlled by a subtropical anticyclone,
ne explanation of which is not properly understood. It occurs both at
round level and aloft, and the subsidence caused by it is responsible
or the extreme aridity of this part of the world. During interglacial
eriods, the intertropical convergence belt, which brings monsoon-type
ains, would have moved farther north in summer than it does today.
Depending on the depth of its penetration, considerable precipitation
ould have reached areas that are now arid or semiarid. Summer rains
vould have fallen in the highlands of Adrar, Air, Tibesti, Ennedi, and,
erhaps, even the Hoggar (Zeuner, 1958).

Evidence for some changes in the climate of the Sahara during the
ast 7000 years is strong. The desert is dotted with upper Paleolithic to
Jeolithic rock engravings as well as the fossil remains of elephant,
hinoceros, hippopotamus, giraffe, and even of domestic animals
vhose present range lies well outside the areas of the petroglyphs
Cooke, 1963). But this climatic change may well not have been so
nfluential as is often thought. I have argued elsewhere (Cloudsley-
Thompson, 1971) that the striking impoverishment of the fauna and
lora, which has occurred since about 3500 B.P., has been due almost
entirely to human activities: bad agriculture, felling trees for fuel, tramp-
ng and overgrazing by domestic stock.

SHIFTING CULTIVATION AND THE FORMATION
OF SAVANNA

Northward from the African rain forest lie three broad belts of savanna
vhose boundaries are by no means clear. Tropical forest grades into
vooded "Guinea savanna," where the rainfall exceeds 100 cm/year,
nearly all of which falls in one period of 7 or 8 months. To the north of
his lies a zone of vegetation known as "Sudan savanna," where the
annual rainfall is in the region of 50 to 100 cm and the dry season lasts
rom October to April. Between that and the Saharan desert steppe, lies
he "Sahel savanna," which enjoys a rainfall of about 15 to 50 cm,
concentrated in 4 or 5 months of the year.

Most African savanna exists in its present form as a result of shifting
cultivation and the use of fire. If left to itself, Guinea savanna, like
'Miombo savanna," its counterpart south of the equator, would probably be thick seasonal woodland or forest, although its floral composi-
ion would naturally differ from evergreen rain forest because it experi-
ences less precipitation and a longer dry season. The climax vegetation
of much of the Sudan and Sahel savanna belts is also probably seasonal
orest, which has likewise been destroyed by man (Cloudsley-
Thompson, 1974).

Destruction of the African forest and the resulting savanna doubtless began when man first acquired the use of fire, probably well over 50,000 years ago. Fire is an essential tool in the shifting cultivation by which so much forest has been destroyed. Although natural fires must occur from time to time, deliberate firing by man has a far greater effect upon the vegetation because man-made fire covers the same ground more frequently and, moreover, is not associated with thunderstorms which may quench the fires that lightning has started.

Fire usually does not spread evenly across whole tracts of vegetation. A mosaic is normally produced as small patches are burned and, later neighboring patches. Some areas may be burned more than once in a year while others escape entirely. Fire favors perennial grasses with underground stems which regenerate readily: in contrast, trees are always damaged, more or less severely. Their trunks become twisted and gnarled, and many of the species that survive best are covered with thick, protective bark. Intermediate types of vegetation are eliminated by constant burning so that a stark contrast is created between wooded savanna, which feeds roaring bush fires, and evergreen forest which is too humid to support a blaze unless it has first been felled and dried. Since the savanna belts of Africa have developed contemporaneously with man (Cloudsley-Thompson, 1969), there is, perhaps, little to be gained in trying to reconstruct their possible floral or faunal composition had man not appeared to make them as they now are. It would be equally mistaken, however, to ignore the human element.

Even in the thickest Sahel savanna, the grasses are sparse and seldom exceed 1 m in height. Consequently, fires here are neither so fierce nor so extensive as in the Sudan savanna belt. Sahel savanna has therefore been reduced to a state from which it is degraded into desert not by burning, but by overgrazing and felling trees for firewood. This has occurred over vast areas of the Sahara within historical times.

OVERGRAZING AND THE CREATION OF DESERT: HISTORICAL EVIDENCE

Two thousand years ago elephants lived among the forests on the foothills of the Atlas mountains: Hanno saw them on the Atlantic shores of Morocco about 500 B.C. Elephants and lions were represented in Meroitic art and were almost certainly captured and tamed in Nubia (Shinnie, 1967). About 60 A.D. the emperor Nero sent an expedition, commanded by two centurians, to discover the source of the Nile. According to Pliny, they found plentiful tracks of elephants and rhinos around Merowe. During the twelfth century A.D., the Arab traveler Idrisi saw elephants and giraffes as far north as Dongola (Jackson, 1957).

In the diary of his journey up the Nile, as late as 1821–1822, Linant de Bellefords mentioned that he heard a lion roaring at Ed Debba near Old

Dongola. In 1835, lions were still plentiful around Shendi, 190 km north of Khartoum, and game was abundant at Kassala as late as 1883. Many other examples have been cited to show how much the flora and fauna of the Sahara has been impoverished in recent times (Cloudsley-Thompson, 1967).

During the period of Roman occupation, the inhabitants of Leptis Magna in Tripolitania were able to draw on the produce of thousands of olive trees. Cereals were cultivated, dates, and even vines, although the climate has probably changed little since then. The land was not fertilized, however, and overexploitation reduced it to desert (Wylie, 1959).

Only 30 years ago, there was dense *Acacia* scrub around Khartoum and Omdurman. This type of plant growth is not now found within 60 miles of the area (Kassas, 1970). Destruction of the trees by felling for fuel has taken place to such an extent that charcoal now has to be brought from as far away as Kosti, Wad Medani, and El Gueisi (Cloudsley-Thompson, 1970). Although the removal of woodland may not have any direct effect on rainfall, as has sometimes been claimed, the presence of trees undoubtedly hinders run-off, enhances the availability of the water to plants, and reduces evaporation. Even more important in creating erosion and desert conditions, however, are overgrazing and compacting of soil by domestic animals, especially goats, which will even climb trees in order to reach leaves. Not only are plants eaten to such an extent that they disappear or die, but regeneration is prevented. Grazing herds cause erosion by loosening the soil, so that it is blown away by the wind; their tracks act as avenues along which water erosion begins, and many plants are killed when their roots are exposed (Fig. 1).

The process of desertification in lands surrounding the Sahara has been reviewed by Kassas (1970). Within living memory, much of Somalia, which is now overgrazed and eroded to a serious extent, was forested, with permanent springs and a rich fauna of elephant and other game (Hemming, 1966). The same is true of the semiarid area around Lake Rudolf.

Wade (1974) cites statements by the Food and Agriculture Organization (FAO) and Agency for International Development (AID) admitting that recent deterioration of the Sahelian zone is due mainly to the fact that modern interventions have usually been narrowly conceived and poorly implemented. Cultivation of cotton and groundnuts have reduced fallow periods from 15 or 20 years to 5 or even 1, and the effect of new boreholes has been to make pasture instead of water the factor limiting cattle numbers so that the inevitable population collapse was all the more ferocious.

Although the present account has been restricted to the Sahara, the desert with which I am most familiar, similar trends are apparent

FIGURE 1 Sahel savanna, showing the effect of severe overgrazing.

throughout most arid regions in the world. Nowhere on earth may sheep or goats safely overgraze, and in the absence of irrigation mechanical agriculture destroys the ecosystem even more rapidly than does overgrazing.

DESERT USE AND RECLAMATION

Arid regions are readily degraded into desert, but the reclamation of their climax vegetation is far less easy to achieve. In any case, most of the world's deserts are increasing rapidly and will continue to do so until the number of domestic animals they support is greatly reduced.

Pastoralism

Of all domestic pastoral animals, the goat is chiefly responsible for enlarging the deserts of Asia and northern Africa. It has largely replaced the migrant gazelle, which only nibbles at thornbushes, and either climbs trees to reach the upper branches or eats them to the ground. Two contrary opinions are commonly expressed about goats. Accordng to the first, these animals are a living testimony to the wisdom of Allah who has created such a wonderful machine that it can transform even waste paper and other refuse into wholesome milk. The second point of view stresses the fact that goats are a menace and there is little hope for the rehabilitation of the Sahara unless they can be eliminated. Personally, I subscribe to the first view in that I do not believe that the desert can be rehabilitated without irrigation, whether goats are present or not. On the other hand, by causing even further degradation of the land, the goat is encouraging the desert to advance. Thus, the future of land that has not yet been reduced to desert is being mortgaged for milk and meat today (Cloudsley-Thompson, 1970). There is no possibility, however, that the number of goats will be reduced voluntarily in the foreseeable future, for these animals represent the sustenance of the people and an insurance for the poor farmer against the years of drought. The combination of overstocking with variability of rainfall leads to considerable fluctuations in the numbers of livestock. For example, during a succession of good years, a livestock population of 8 million head was built up in Algeria. After a few years of drought, however, it was reduced in 1945 to 2 million (Droughin, 1962).

Adding to the number of points at which livestock can secure water helps to widen the area of grazing and reduces pressure around the original sources of water. It can be achieved by boring new wells and by constructing *haffirs* or artificial ponds in which seasonal rainwater may be stored. Such measures can be effective, however, only so long as the herds are not permitted to increase. If they do, the situation becomes worse and not better. Where these flocks and herds represent material wealth and the population is almost entirely dependent upon them, both as food and currency, it is not yet possible to counter the rape of the earth by overgrazing (Cloudsley-Thompson, 1970). Pastoralism is clearly a major cause of desert expansion and cannot therefore be advocated as a form of land use in arid regions, except under strict control and supervision.

Nomadism

Present solutions to the problem of desert expansion depend upon drastic policies such as the construction of large dams and irrigation schemes, that change the entire economy and way of life of the inhabitants. Government policies toward nomadism are equally unenlightened. They appear to be directed chiefly toward the settlement of the nomads and the restriction of their migration routes, although the nomadic way of life is probably the only one which will ever produce much in the way of food from deserts, with their erratic and unpredictable rainfall. Traditional nomadism is a remarkably efficient and ecological adaptation to the desert environment (Wade, 1974). If it were to disappear, vast areas that are now productive would become permanently useless to mankind. It might be better in the long run, therefore to encourage and modernize the nomadic way of life. The hardship that nomadic people have to endure could be ameliorated by a flying doctor service, mobile markets, and practical educational facilities Grazing could be controlled and improved, news of distant rainfall transmitted by radio, and so on.

Unlike cattle, sheep, and goats, camels can withstand a high degree of dehydration and still maintain their appetites. Consequently, their grazing areas, which are proportional to the number of days that they can go without water, are very much greater than those of other domestic animals. For this reason, the camel offers the most promising solution to problems of increasing meat production in arid areas where there is a low density of vegetation that cannot easily be increased (Schmidt-Nielsen, 1956). There is little point in expanding the production of camels, however, unless they can be exported or local people persuaded to eat them. Development plans aimed at increasing animal production must take into account the preferences of the population they are intended to benefit. It is worth considering whether camel mea could be preserved satisfactorily as dried biltong (Cloudsley-Thompson and Cloudsley-Thompson, 1970).

Game Ranching

Another way of utilizing available resources more effectively might be to employ the principles of game ranching, whose value has been so effectively demonstrated in other parts of Africa (Cloudsley-Thompson 1967). Gazelles, addax, oryx, gerenuk, and other antelope provide excellent meat and are well adapted to life in arid regions. The mobility of these animals, upon which their survival depends, at the same time makes them difficult subjects for ranching. Like camels, certain East African species, such as buffalo, beisa oryx, eland, and black rhinoceros, require relatively little water. They can endure comparatively large variations in body temperature without sweating and thus conserve moisture (Cloudsley-Thompson, 1969). As many game animals once ranged into quite arid regions, it might be possible one day to consider game ranching with species now restricted to the wetter parts of Africa. Ostriches too could easily be ranched in the Sahara to provide meat and feathers, if poaching and unauthorized hunting were controlled. But this would be almost impossible at the moment because of the vast areas involved and the unenlightened attitude of the people toward wild animals.

In such circumstances, game ranching cannot be instituted without considerable research and capital outlay. Airplanes, helicopters, and landing strips would be required to observe poachers, locate game, and transport staff and supplies. Unless scientific and financial aid is supplied, such schemes will never develop (Cloudsley-Thompson and Cloudsley-Thompson, 1970).

Utilization of Natural Desert Plants

Cactuses and other desert succulents can be used by man as a source of moisture, as are gourds and *tsama* melons by the Bushmen of the Namib. In Texas, the thorns of yucca and prickly pear (*Opuntia inermis*) have been burned off with petrol torches in times of drought to provide food and water for famished cattle. The mouth of the camel is so tough and leathery, however, that no such treatment is necessary to make prickly pear acceptable to him. Spineless cacti (*Opuntia ficus-indica*) are an important food of livestock in North Africa, and sheep can be fed on them exclusively for several weeks without suffering any ill effects.

Although prickly pear thrives in Eritrea and on the Mediterranean coast, it is not found extensively along the southern fringe of the Sahara, probably because climatic conditions are there too extreme. But the possibility of its introduction into new areas certainly merits investigation. If it became established, not only would it provide food for domestic animals, but it might also help to check extensive run-off from rain storms (Cloudsley-Thompson, 1970).

In addition to its importance as cover and feed for domestic animals, desert vegetation includes some food, beverage, and medicinal plants, and sources of oils, waxes, and fibers. In general, the harvest of wild desert species is centered on arid areas that have low carrying capacity for grazing and insufficient rainfall for dry farming. Most of these natural products, however, have been replaced by synthetic substances. The prospect for economic harvesting of natural supplies is therefore poor. Certain xerophytic plants, such as *Agave sisalona* and *A. fourcroydes,* which produce the fibers sisal and henequen, respectively, can nevertheless be used to extend cultivation into semiarid and arid regions in which ordinary cultivated crops fail because of drought and grazing gives a low return. Although the outlook for such cultivated arid-zone plants is considerably better than it is for wild species, exploitation still depends greatly upon cheap labor. With an increasing world demand for raw materials, however, the products may, perhaps, be able to maintain their value despite competition from synthetic materials (Duisberg and Hay, 1971).

Irrigation Schemes

The most obvious method of increasing productivity of arid lands is by irrigation. Where water is available and soils are suitable, deserts can be transformed into green and productive areas. Much of the underground water of the Sahara is saline, however. The alkaline carbonates and bicarbonates present render it unacceptable to crops, and sulfates of sodium and magnesium have purgative properties and affect the use of the water by man and domestic animals. Present methods of desalinization are costly and require skilled supervision. They could not, therefore, presently be introduced into much of the Sahara, even if it were possible to increase the production of underground water supplies—and prospects of this are far from encouraging. Moreover, until people who irrigate their crops are organized, such projects will not be used effectively. Even more impracticable at present are schemes such as that to transport icebergs from the Antarctic to desert areas of the world.

More efficient use of existing water can be made through hydroponic cultivation. Nutrient solution, pumped twice daily through plastic tubes perforated by small holes, will irrigate plant roots without wetting the surface of the soil, so that little water is wasted through evaporation. In this way, for example, 1 ton of tomatoes can be produced from 40 to 50 m^3 of water instead of 80 to 150 m^3 required by plants grown with surface irrigation.

Where irrigation water is abundant, plants grow rapidly in the desert provided that the soil is suitable. When radiation, temperature, and rainfall, are known, water requirements and productivity of agricultural crops can be predicted. Thus, at least 50 percent greater growth rate is to be expected from sunny savanna and desert areas than from regions of cloudy rain forest. Up to a dozen crops of alfalfa per year or four of wheat and barley have been gathered in parts of Libya where water can be pumped from a depth of 40 m.

However, large-scale desert irrigation schemes are almost certain to become focal centers for bilharzia, malaria, and other diseases transmitted by invertebrate vectors. At the same time, the crops they produce are particularly vulnerable to attack by desert locusts (*Schistocerca gregaria*) and other insect pests, which are provided throughout the year with conditions suitable for breeding (Cloudsley-Thompson, 1970).

In this context, however, it may be worth mentioning that, although the efficiency of present surveys is variable and usually low, the detection of potential breeding sites of the desert locust by remote sensing from satellites offers hope of a detailed routine survey (Pedgley, 1972). No doubt, advanced technology will eventually be applied to other biological problems in arid regions, but it should not take the place of ecologically based land use.

Insect pests have a greater impact on the economies of underdeveloped nations than on those of most Western countries, mainly because insects breed faster in the tropics than in temperate regions. More and more pests are becoming resistant to an increasing number of chemicals and natural enemies are increasingly reduced. In addition, the accumulation of insecticides in human tissue is greater in countries of the "Third World" and its effect more serious where nutritional levels of the population are low (Zethner, 1973).

Dryland Farming

Where there is no source of irrigation water, however, and precipitation is limited to a short wet season, annual crops tend to be much more productive than perennial plants. Many semiarid regions are therefore exploited for the cultivation of cereals and other annual plants. Traditional dry farming methods that do not overexploit the soil are widely practiced and provide the sustenance for many sedentary desert people. Large-scale dryland farming schemes that employ mechanical ploughing are potentially extremely dangerous and may well induce considerable soil erosion. The new Gedaref dura (*Sorghum dura*) project, for example, which is introduced into a fragile, semiarid environment of the Sudan, is likely, in the long run, to prove a costly failure from which only locusts and grasshoppers will benefit. Small areas had previously been farmed traditionally with ox ploughs and left fallow for several years after harvesting. The new intensive scheme, financed by the World Bank, does not appear to have taken such ecological principles into consideration.

Stabilization of Dunes

In areas where rainfall exceeds 150 mm a year, sand dunes may be capable of supporting permanent vegetation. Even at the height of the dry season in such areas the dune sand is moist a short distance below the surface, the rainwater that enters the dune being stored from one wet season to the next. Little capillary rise of moisture above the water table occurs and the dry sand on the surface protects the water in the deeper layers from evaporation. However, if wind removes the surface layers, the moist underlying layers are exposed and evaporation is rapid. Of course, such winds will also uncover the roots of plants, subjecting them to dehydration and killing them. Further, the wind-blown sand itself will act as a sandblast and destroy aboveground vegetation. Therefore, if the surface layers of sand dunes are stabilized, the chances of establishing plant cover are good. Stabilization may be achieved by spraying the dunes with a mixture of oil and synthetic rubber, before planting seedlings of *Acacia, Eucalyptus,* etc. Once vegetation is established, the dune surface becomes permanently stabile, first, because as the plants grow they provide a natural windbreak that reduces the wind velocity near the dune surface and, second, because dead foliage becomes incorporated into the soil and increases its cohesiveness.

DISCUSSION

Various solutions have been proposed to the problems of desert expansion and its reclamation. The long-term scientific answer, both in rich and in poor countries, is not difficult to find. It lies, of course, in multiple land use, some applications of which have been described here; traditional land use in arid zones is discussed by Arnon (1972).

Although much could still be done for semiarid and desert lands of the world, it is well to be realistic. Although it might be more profitable to invest in agricultural land that is already productive than to attempt to develop marginal areas, this would not help their inhabitants to help themselves. At the same time, much of the money and technical aid supplied to developing countries by national and international agencies is misapplied or wasted. To help the people of the desert, one must understand and sympathize with their traditions, religious attitudes, problems, and ways of life. Small improvements that are immediately effective may be more valuable than grandiose schemes, which often come to nothing. Development plans must be accompanied by education, and the best projects rely less on money than on realizing the latent abilities of the people that they are intended to help. Desert people are intelligent and individualistic. They value their traditions and, although they may appreciate help, they naturally dislike a patronizing attitude toward their problems. Educators, administrators, and officials need, therefore, to be both sympathetic and understanding.

It is gradually becoming better known that increased yields are not ends in themselves. Development for a country as a whole takes place only when increased production is distributed evenly among the population. Western technical innovations often fail or have adverse effects on the societies into which they are introduced (Zethner, 1973).

Except where large quantities of water are available for irrigation, the agricultural productivity of desert can never be great. Some protein can be produced by the herds of the local people, but probably at the cost of further expansion of the desert. In any case, since agricultural societies require more water per head of population than do industrial com-

unities, it is worth considering whether alternative uses of arid
regions—for mining, industry, solar energy, tourism, etc.—might not be
more promising in the long run, especially in view of the high cost of
producing new water (Wollman, 1970). In the immediate future, how-
ver, there seem to be few practical ways of developing arid regions
unless oil, water, or other subterranean resources are present. In any
ase, a multiplicity of types of land use will press less heavily on the
nvironment than single usage.

Many of the world's mineral reserves, upon which Western civilization
ow depends, are located in developing countries. There is hope that
through cooperation some redistribution of wealth may result in improved standards of living in these poor, arid lands. This might be
chieved only at the cost of less extravagance, pollution, and waste in
the affluent Western Societies. If, and when, this happens, desert people
ill not be so dependent upon their stock, and the desert may at last
nally cease to expand. At the same time, the quality of life may improve
mong the industrial nations of today.

LITERATURE CITED

rnon, I. 1972. Crop production in dry regions: Vol. 1, Background and princi-
    ples; Vol. 2, Systematic treatment of the principal crops. Leonard Hill,
    London.

akker, E. M. van Zinderen. 1963. Palaeobotanical studies. S. Afr. J. Sci.
    59:332–340.

utzer, K. W. 1959. Studien zum vor und früchgeschichtlichen Land-
    schaftswadel der Sahara III. Die Naurlandschaft Ägyptiens während der
    vorgeschichte und der Dynastichen Zeit. Abh. Math. Naturw. Kl., Akad. Wiss.
    Mainz 2:44–122.

loudsley-Thompson, J. L. 1967. Animal twilight: man and game in eastern
    Africa. Foulis, London.

———. 1969. The zoology of tropical Africa. Weidenfeld & Nicolson, London.

———. 1970. Animal utilization. In H. E. Dregne (ed.), Arid lands in transition,
    pp. 57–72. American Association for the Advancement of Science, Washing-
    ton, D.C.

———. 1971. Recent expansion of the Sahara. Intern. J. Environmental Studies
    2:35–39.

———. 1974. The expanding Sahara. Environmental Conserv. 1:5–13.

———, and J. A. Cloudsley-Thompson. 1970. Prospects for arid lands. New
    Scientist 48:286–288.

ooke, H. B. S. 1963. Pleistocene mammal faunas of Africa, with particular
    reference to southern Africa. In F. C. Howell and F. Boulire (eds.), African
    ecology and human evolution, pp. 65–116. Methuen, London.

roughin, G. 1962. The possibility of using minor water resources in Algeria.
    Arid Zone Res. 18:371–380.

uisberg, P. C., and J. L. Hay. 1971. Economic botany or arid regions. In W. G.

McGinnies, B. J. Goldman, and P. Paylore (eds.), Food, fiber and the arid lands, pp. 247–270. University of Arizona Press, Tucson, Ariz.

Flohn, H., and M. Ketata. 1971. Étude des conditions climatiques de l'avance d' Sahara tunisien. W. M. O. Tech. Note, 116:1–32.

Grove, A. T. 1974. Desertification in the African environment. African Affairs 73:137–151.

Hemming, C. F. 1966. The vegetation of the northern region of the Somal Republic. Proc. Linn. Soc. Lond. 177:173–250.

Jackson, J. K. 1957. Changes in the climate and vegetation of the Sudan. Sudan Notes Rec. 38:47–66.

Kassas, M. 1970. Desertification versus potential for recovery in circum-Sahara territories. In H. E. Dregne (ed.), Arid lands in transition, pp. 123–142 American Association for the Advancement of Science, Washington, D.C.

Kendall, R. L. 1969. An ecological study of the Lake Victoria basin. Eco Monogr. 39:121–176.

Pedgley, D. E. 1972. Satellite detection of potential breeding sites of the deser locust. Proc. Symp. S. 62 Remote Sensing, Pretoria CSIR, 155–158.

Schmidt-Nielsen, K. 1956. Animals and arid conditions: physiological aspects o productivity and management. In G. F. White (ed.), The future of arid lands pp. 368–382. American Association for the Advancement of Science, Wash ington, D.C.

Shinnie, P. L. 1967. Meroe. Thomas & Hudson, London.

Wade, N. 1974. Sahelian drought: no victory for Western aid. Science, 185:234 237.

Warren, A. 1970. Dune trends and their implications in the central Sudan. Z Geomorph., Suppl. 10:154–186.

Winstanley, D. 1973. Rainfall patterns and general atmospheric circulation Nature, 245:190–194.

Wollman, N. 1970. Economics of land and water use. In H. E. Dregne (ed.), Ari lands in transition. pp. 143–163. American Association for the Advancemen of Science, Washington, D.C.

Wylie, J. C. 1959. The wastes of civilization. Faber & Faber, London.

Zethner, O. 1973. Assistance to agricultural scientists in the third world: neglected field. World Development 1:53–55.

Zeuner, F. E. 1958. Dating the past, 4th ed. Methuen, London.

ENVIRONMENTAL PHYSIOLOGY OF DESERT ORGANISMS: SYNTHESIS AND COMMENTS ON FUTURE RESEARCH

Neil F. Hadley

Department of Zoology, Arizona State University, Tempe, Arizona

Symposium participants have provided a detailed account of the physical conditions associated with deserts and the primary adaptations possessed by a variety of organisms which allow them to either tolerate or avoid climatic extremes. Coverage includes behavioral and physiological adaptive patterns for the plant or animal as a whole, adaptive mechanisms and processes at the cellular and molecular levels, and man's activities in arid regions and their influence on desert ecosystems. Neither space nor practicality permit a thorough synthesis of these complex and voluminous data. Hence, I have chosen to highlight three main themes that provide the foundation for the symposium and appear throughout these contributed papers: (1) similarities in the mechanisms used by plants and animals to solve the problems of desert living, (2) the development and utilization of instrumentation and techniques that enable acquisition of realistic and meaningful data, and (3) specific research areas and problems that are primary targets of future investigations.

ADAPTATIONAL SIMILARITIES IN PLANTS AND ANIMALS

Most desert organisms are subjected to excessive heat and dehydration during at least some part of their life histories. Even those species which can escape the climatic extremes through modification of activity cycles typically exhibit a combination of morphological, physiological, and biochemical adaptations that make them better adapted than their counterparts inhabiting more mesic habitats. A priori one might expect that the nature of the adaptive mechanisms exhibited by plants and animals would be as different as the directions evolutionary processes have taken in these two groups. However, a closer examination reveals just the opposite to be true—that plants and animals in arid environments show many striking similarities in their morphological and physiological adaptations.

An analysis of adaptive mechanisms common to both life forms was presented by Hadley (1972). The survey stressed similarities shared by desert cacti and xeric-adapted arthropods, although references were also made to adaptive features common to plants and higher vertebrates. Similarities observed included: (1) tolerance of high temperatures; (2) reduction in incident thermal load through changes in orientation and the presence of reflective–absorptive surface projections; (3) increased thermolability (heterothermy); (4) presence of lipid layers to reduce transpiration; (5) modification of the location, aperture, and control of gas-exchange openings; (6) mechanisms for economically transporting and eliminating excess salts and electrolytes; (7) absorption and utilization of atmospheric and/or substrate moisture; (8) a greater, and in some cases, total dependence upon metabolic water; and (9) tolerance of extreme cell and tissue desiccation.

A number of examples that support and extend these observations have been presented by the symposium authors. Ability to not only tolerate but function at high temperatures is perhaps the one adaptive mechanism most commonly observed in both life forms. Adjustments at the cellular and subcellular level provide this basis and occur in a variety of physiological processes. Examples include enhanced enzyme–substrate affinities and increased thermal stability of membrane molecules that enable photosynthesis and respiration to continue at high temperatures or permit continued active transport of a critical electrolyte. Detailed investigations of enzyme kinetics, metabolic pathway controls, and molecular stability in desert-adapted organisms are needed to elucidate the exact nature of these adjustments.

Mechanisms for reducing the heat load and/or dissipating heat gained are also frequently exhibited by desert plants and animals. Although the most effective of these, behavioral avoidance, is essentially restricted to animals, both groups utilize orientation to either increase or decrease the amount of radiation absorbed (e.g., positioning of pads in succulents, postural adjustments in animals).

Surface projections and irregularities absorb and reflect much of the incoming solar radiation and thus help prevent tissues from reaching intolerable temperatures. They also contribute to the formation of an effective "boundary layer" that is instrumental in determining the rate and quantity of heat transfer between the organism and its immediate environment. In plants these morphological projections take the form of spines, trichomes, and stem convolutions; analogous structures in animals include setae, epidermal tubercles, scales, and hairs. Apparently the increased boundary layer thickness and reflectance created by these projections more than compensate for the increased surface area.

Surface coloration and texture complement the effects of morphological projections in modifying thermal flux. Under conditions when heat is detrimental to an organism, white or light coloration will reduce the rate of heat gain in comparison to black or dark forms. Energy-flow differences due to coloration are better documented in animals (Norris, 1967; Ohmart and Lasiewski, 1971; Zervanos and Hadley, 1973) than plants, which are perhaps more restricted because of photosynthetic limitations. Nevertheless, much of the reflection of incident solar radiation by the cactus *Cephalocereus senilis* results from the almost "white" appearance created by the hairlike spines. Reflectivity and absorption produced by less conspicuous features such as "tubercles" on the elytra of beetles, rough-textured wax layers on the surface of succulents, or accumulation of NaCl crystals in dehydrated leaves of *Atriplex* may also play an important role.

Convergent parallels are also abundant when one examines the adaptive responses of plants and animals to the other basic problem encountered—dehydration. In many cases the modifications for water conservation are much the same as those for withstanding temperature stress (Treshow, 1970). Two features stand out from the point of effectiveness and similar design. These are the presence of integumentary waterproofing barriers, and mechanisms for restricting water loss during gas exchange.

One of the most effective adaptations for living in desiccating environments is having an integument that is relatively impermeable to water. The nature of this restrictive barrier may take several forms. For many species just the inherent structural arrangement of integumentary components provides resistance to water flux. Superficial structures, such as scales, horny layers, or thickened keratinized epidermis, are all suspected of being effective in reducing water loss through the integument, although in many cases this assessment is based upon "appearance" rather than quantitative supporting data. Recent investigations have questioned the functional significance of the above structures and suggest that some other integumentary feature or process is responsible for this barrier function (Licht and Bennett, 1972; Hattingh, 1972). Other "membranes" that deserve further study in terms of their role, if any, in water balance include the egg coverings and ephippia of aquatic crustaceans, cocoons formed during estivation of some desert amphibians, and outer layers of eggs laid by oviparous desert reptiles.

The low transpiration rates that characterize most desert-adapted plants (particularly succulents and certain shrubs) and arthropods (beetles, scorpions, spiders) reflect the presence of a thick epicuticular wax, although in both groups other integumentary layers and components probably contribute to cuticular resistance. The structural arrangement and chemical composition of these layers are complex and highly variable (see reviews by Jackson and Baker, 1970; Eglinton and Hamilton, 1967). Possible causes of the observed variation include sex differences, diet, life stage, and the immediate necessity for water conservation. Few studies have tested these possibilities, nor have correlations been sought between chemical composition and environmental conditions. Especially valuable would be investigations of structural and compositional changes that occur at the transition temperature, correlation of lipid composition with thermal regimes experienced and degree of water impermeability, and determination of sites and pathways for the biosynthesis of cuticular waxes.

Although a lipid layer effectively reduces cuticular transpiration, it also restricts oxygen and carbon dioxide exchange. To circumvent this problem, both plants and arthropods evolved respiratory systems in which external openings are restricted to small pores on the surface. The striking morphophysiological similarities between trachea-spiracles of arthropods and plant stomates are illustrated and discussed in detail by Hadley (1972). These respiratory systems permit gas exchange coincident with effective cuticular resistance to water flux. Their component structures have been modified or their microenvironment altered so that respiratory transpiration also has been markedly reduced. Openings that lead to internal gas-exchange chambers in xeric-adapted arthropods and plants are characteristically sunken and often superficially covered by scales or hairs. Such an arrangement increases diffusion resistance by lengthening the diffusion pathway and increasing boundary-layer resistance. When water supply dictates further conservation, openings to these structures can be completely closed in both groups. Cyclic CO_2 release patterns exhibited by diapusing stages of insects and the crassulacean acid metabolism pathway for CO_2 fixation in desert succulents are consequences of pore aperture regulation.

The most successful desert species usually possess a combination of effective adaptations to cope with physical and biotic extremes. Desert scorpions are excellent examples of such a group (Hadley, 1974). Lichens and aquatic crustaceans are champions, however, when temperature tolerance and desiccation resistance are considered. Symposium participants Lange and Belk/Cole highlighted these two groups, which, despite their obvious differences, show parallel responses in the nature of their adaptive mechanisms. Dehydrated lichens are essentially immune from either heat or cold injury in their native habitat. Likewise, certain aquatic insect larvae tolerate water temperatures of 60°C, while lethal temperatures for cryptobiotic phyllopod eggs are well above this extreme. Both lichens and aquatic crustaceans (egg stages) are capable of withstanding nearly total dehydration. In fact, application of the term "arido-passive" to desert lichens could be extended to many desert temporary pond inhabitants, since both groups limit their metabolic activities to moist periods and become inactive and dehydrated during stress periods.

Reproductive strategies also play an integral role in the adaptive biology of these and other desert-adapted species. The extremely tolerant egg stage of crustaceans inhabiting temporary ponds has as its counterpart the temperature-insensitive seeds produced by desert annuals, another group of "arido-passive" plants. Reproductive effort appears to be very high in these two groups. Prolific seed and egg production is necessary to ensure that some offspring survive periods of drought and high temperatures to replenish population numbers during the more favorable conditions that follow. Sufficient numbers are needed to compensate for those eggs and seeds that err in interpreting environmental clues for terminating diapause or initiating germination and are unable to complete their growth and reproduction. Quantitative experimental determinations of the portion of total energy expenditure allocated to reproduction by xeric-adapted organisms may provide valuable clues to understanding the total adaptive process (Tinkle and Hadley, 1975).

A final point in this comparison is the ability to absorb and utilize either atmospheric and/or substrate moisture. This phenomenon is exhibited by both desert plants and animals, although the relative importance of this avenue of water gain varies greatly. Most desert animals obtain sufficient water from surface supplies, their prey, or metabolism of foodstuffs to balance that lost through transpiration and elimination of nitrogenous end products. Exceptions here are a few desert arthropods (Edney, 1966, 1971; Louw, 1972), and especially desert amphibians which rely on integumental uptake of moisture resulting from summer showers. Similarly, plants depend primarily on extensive penetration of root systems and water storage by these and other organs, although there is increasing evidence that even higher plants located in coastal desert regions can at least supplement their moisture needs by utilization of condensate from advective fogs. For desert lichens atmospheric water vapor and dew represent the only water sources necessary for metabolic processes.

RESEARCH DESIGN—NEW APPROACHES TO PROBLEM SOLVING

Environmental physiology, like all areas of science, has benefited greatly from the development and application of sophisticated instrumentation and innovative techniques to provide basic and meaningful data. Reports from symposium participants indicate that current desert studies are no exception. The use of scanning electron microscopes to elucidate surface projections that are effective in radiation absorbance and reflectance, fine thermocouples, thermistors, and infrared radiometers to determine surface and tissue temperatures, analytical instrumentation for the establishment of physical and chemical properties of microsamples, chromatographic and fluorometric techniques for the identification and isolation of important chemical compounds, and the employment of a variety of labeled substances to determine uptake rates, carbon budgets, hormonal activity, and enzyme pathways are but a few examples of such applications discussed.

An important objective reflected in several presentations and one that should be an inherent feature of any ecophysiological investigation of desert organisms if possible is the desirability to obtain information on nonrestricted animals under natural desert conditions. The extension of radioisotopic and telemetric techniques has made this possible for a variety of organisms. Included are determinations of uptake and elimination rates of water and/or critical electrolytes, measuring field metabolic rates, tracing the flow and allocation of carbon by plants, and constructing energy-exchange budgets for free-roaming animals. Complementing this approach has been the use of mobile laboratories to bring instrumentation and data-acquisition systems to the species' habitat or the simulation of actual habitat conditions by elaborate laboratory environmental chambers and phytotrons. Not only does this type of investigation counter some of the difficulties created by artificial laboratory conditions, but it also provides a validity check on previously gathered laboratory data and can be used to determine if less time-consuming and inexpensive indirect methods might be more profitably employed. It is likely that as more studies of this nature are conducted, critical data on the techniques and methods employed as well as on the physiological processes being measured will be forthcoming.

FUTURE RESEARCH

Despite the numerous studies conducted on xeric-adapted organisms and the voluminous literature that has resulted, our knowledge of the adaptive biology of these species is far from complete. Future investigations are likely to continue to shift away from measurements of degree of tolerance or rate functions of organisms under stress conditions to experiments to uncover the biochemical and genetic bases for physiological adjustments and the mechanisms that govern their control. Comparative studies, however, are still needed at all levels, especially where conclusions and theories are based upon data gathered from a limited number of taxon representatives. These new data, plus reevaluation of existing data in light of new techniques, may lead to modification of many previously accepted facts and the total scrapping of others.

Future research in all areas should stress the continued integration of scientific disciplines such as radiochemistry and biophysics, and employ quantitative models that predict behavioral and physiological parameters as framework for further experimental testing. Wherever possible, field-oriented studies on species under "natural" conditions should be attempted to confirm laboratory findings and provide the investigator with ecologically significant data.

Finally, I feel there is a critical need for more "applied" research by environmental physiologists, especially those working in desert ecosystems. We can and must utilize information accumulated from previous studies as well as design future experiments to provide answers to current problems. With the continuing rapid growth of the world's population and decreasing food supply, man will become increasingly dependent upon arid regions for living space and particularly for food production. Providing information on how to live and work more effectively and comfortably in a hot, dry environment is an important but only small part of our potential contribution. Emphasis should be placed on the development and testing of plant strains that not only survive but that have a high productivity under such conditions. Similarly, we must apply our understanding of thermal control mechanisms to better develop the role of domestic species in these habitats (Bligh, 1972). Indirect, but equally effective, methods for increasing food supply are also available from environmental physiology investigations of thermal tolerances and water relations of insect pests in agricultural regions. Using alfalfa as a test crop, Pinter et al. (1975) demonstrated the potential value of cultural manipulation of the environment (i.e., strategically timed cutting, rapid repositioning and baling of dry windrow vegetation, delayed first irrigation) to expose sheltered insects to extreme soil and air temperatures. Such biological control measures would be economically sound and have minimal negative environmental impact.

LITERATURE CITED

Bligh, J. 1972. Evaporative heat loss in hot arid environments. In G. M. O. Maloiy (ed.), Comparative physiology of desert animals, pp. 357–369. Academic Press, New York.

Edney, E. B. 1966. Absorption of water vapour from unsaturated air by *Arenivaga* sp. (Polyphagidae, Dictyoptera). Comp. Biochem. Physiol. 19:387–408.

———. 1971. Some aspects of water balance in tenebrionid beetles and a thysanuran from the Namib Desert of southern Africa. Physiol. Zoöl. 44:61–76.

Eglinton, G., and R. J. Hamilton. 1967. Leaf epicuticular waxes. Science 156:1322–1334.

Hadley, N. F. 1972. Desert species and adaptation. Amer. Sci. 60:338–347.

———. 1974. Adaptational biology of desert scorpions. J. Arach. 2(1):11–23.

Hattingh, J. 1972. The correlation between transepidermal water loss and the thickness of epidermal components. Comp. Biochem. Physiol. 43A:719–722.

Jackson, L. L., and G. L. Baker. 1970. Cuticular lipids of insects. Lipids 5:239–246.

Licht, P., and A. F. Bennett. 1972. A scaleless snake: tests of the role of reptilian scales in water loss and heat transfer. Copeia (1972):702–707.

Louw, G. N. 1972. The role of advective fog in the water economy of certain Namib Desert animals. In G. M. O. Maloiy (ed.), Comparative physiology of desert animals, pp. 297–314. Academic Press, New York.

Norris, K. S. 1967. Color adaptation in desert reptiles and its thermal relationships. In W. W. Milstead (ed.), Lizard ecology: a symposium, pp. 162–229. University of Missouri Press, Columbia, Mo.

Ohmart, R. D., and R. C. Lasiewski. 1971. Roadrunners: energy conservation by hypothermia and absorption of sunlight. Science 172:67–69.

Pinter, P. J., N. F. Hadley, and J. H. Lindsay. 1975. Alfalfa crop micrometeorology and its relation to insect pest biology and control. Environ. Entomol. 4(1):153–162.

Tinkle, D. W., and N. F. Hadley. 1975. Lizard reproductive effort: caloric estimates and comments on its evolution. Ecology (in press).

Treshow, M. 1970. Environment and plant response. McGraw-Hill, New York. 422 p.

Zervanos, S. M., and N. F. Hadley. 1973. Adaptational biology and energy relationships of the collared peccary (*Tayassu tajacu*). Ecology 54:759–774.

INDEX